BEAUTY WITHOUT THE BEASTS

See the beauty in all beings. Drawing by Amanda Sehic

BEAUTY WITHOUT THE BEASTS
A Guide to Cruelty-Free Personal Care

Heather Chase

Lantern Books • New York
A Division of Booklight Inc.

2001
Lantern Books
One Union Square West, Suite 201
New York, NY 10003

Copyright © Heather Chase 2001

All rights reserved. No part of this book may be reproduced, stored in a retrieval system, or transmitted in any form or by any means, electronic, mechanical, photocopying, recording, or otherwise, without the written permission of Lantern Books.

Hands
Words and music by Patrick Leonard and Jewel Kilcher. © 1998 EMI APRIL MUSIC INC., BUMYAMAKI MUSIC and JEWEL KILCHER. All Rights for BUMYAMAKI MUSIC Controlled and Administered by EMI APRIL MUSIC INC. All Rights Reserved International Copyright Secured Used by Permission.

Printed in the United States of America

Library of Congress Cataloging-in-Publication Data

Chase, Heather
    Beauty without the beasts : a guide to cruelty-free personal care / Heather Chase
        p. cm.
    Includes bibliographical references.
    ISBN 1-930051-60-3
        1. Beauty, Personal. 2. Animal experimentation. I. Title

RA778. C395 2001
646.7—dc21

2001038048

I dedicate this book to all souls who share this home we call Earth. May the qualities of respect, compassion and love grow within each of us, so that we may live together in harmony.

Table of Contents

Acknowledgments 9
1: My Journey 11
2: Models with Conscience 21
3: Aware Personal Care 35
4: Compassion in Fashion 67
5: Ethical Eating 83
6: More Ways to be Cruelty-Free 99
7: Cosmic Connections 115
Resources 125

Acknowledgments

I am deeply grateful to God for blessing me with the precious gift of life. Also, I appreciate the many animals who have enriched my life with companionship, insight, laughter, and unconditional love. I thank my parents, Ron and Janet Chase, for their constant support and for encouraging me to question the status quo, to believe in my dreams, and to make a unique contribution to the world. Thank you to my dear friend, Rob Robb, for his profound wisdom, generosity, humor and for helping me let my true colors shine. Countless other people and organizations have generously supported this project. The major ones are listed in the Resources section and their assistance is profoundly appreciated. Finally, I thank you, the reader, for your interest in this book. Enjoy!

Welcome to my world! Photo by Steve Thompson.

1

My Journey

The only lasting beauty is the beauty of the heart. —Rumi

I ADMIRE BEAUTY AND I LOVE ANIMALS. THAT'S why I'm writing this book—to help you enhance your beauty without harming animals. *Beauty without the Beasts* will reveal the suffering that may be lurking in products you use every day, and will offer information, tools, and resources to help you choose items that do not involve such misery.

In researching the topics addressed in this book, I've seen extremely gruesome photographs revealing the wretched results of some people's cruelty to animals. Viewing those pictures turned my stomach and saddened my heart. I trust that you share a basic sensitivity to other living beings. My aim in *Beauty without the Beasts* is to reveal the truth without disgusting you so much that you slam the book shut and never open it again. So, I am not going to subject you to any repulsive photos of suffering animals. Instead, I will use words and let your imagination and heart do the rest. By using cruelty-free products and living in tune with our individual conscience, each of us may naturally radiate even more of our personal beauty. It's good for us *and* the animals!

First, I'd like to tell you a little about myself, and share with you the journey that has led me to write this book. As an only child, I have enjoyed the companionship of animal "siblings" throughout my life. I used to spend hours playing with my animal friends—even dressing them up in doll clothes (which they patiently tolerated). When I felt sad and alone, their soothing presence gave me a solace I could not find elsewhere. At my elementary school, several stray animals visited the playground and, every day during recess, I'd give the stray dogs part of my lunch. Several times I hid stray cats under my jacket, smuggled them onto the school bus, and brought them home to live at my house. I did this so often that at one time about six different cats were living there!

Me snuggling a guinea pig at the San Francisco SPCA. Jacket with faux fur collar by Esprit. Photo by Russell Tanoue, make-up by Jessica Campo, hair by Danica Winters.

Having such a close relationship with animals, I was bewildered and horrified whenever I saw people hurting them. One of my saddest childhood memories is of me looking through my bedroom window at a small, brown dog who was chained up in the neighbor's yard. When the neighbor came home that day, the dog barked and jumped around excitedly. Seeing the dog's joy made me smile. But the man didn't share our happiness. Instead, for some inconceivable reason, he beat the dog with a rolled-up newspaper. The little dog looked up with a perplexed expression in his eyes as he cowered and whimpered. Watching this sad turn of events, my heart went out to the innocent dog, but I was afraid to help him. After all, the man beat his own dog for no reason—what would he do to me if I confronted him? So, I just let the tears stream down my face as I watched through the window's sheer curtain.

This memory came back to me during my early twenties, when, after earning a Bachelor of Fine Arts degree and working as a secretary, I did a lot of soul-searching in hopes of finding my true "purpose" in life. A book called *Zen and the Art of Making a Living*, by Laurence G. Boldt, prompted me to reflect on what I most loved as a child and what childhood memories most stirred my emotions. It was clear to me that I would enjoy working with or helping animals in some way.

Around the same time, I met a remarkable clairvoyant named Rob Robb, who gave me profound insight into my aura colors (yellow and violet), my personality, and my potential. He told me that I have the ability to inspire people simply by virtue of my presence, and that it was my "job" in life to do this. Rob encouraged me to explore career options, such as acting or modeling, in which I could express myself and radiate in front of many people. I had done some modeling in the past, but was hesitant to pursue it full-time because I saw it as a superficial career, concerned only with vanity and money. Plus, I was shy and reluctant to draw too much attention to myself. I kept Rob's advice in the back of my mind as I continued to work and repay my student loans.

Each member of my family, including myself, was at that time working in a stressful, unsatisfying job. Like many people whose work lacks passion and true meaning, we lived for the weekends and dreaded going to work in the mornings. We finally decided to resign from those jobs and take time off to find our true life's work by going on a sojourn throughout the Western United States.

During this journey, I researched animal-related careers and came across information about the practice of testing cosmetic ingredients on animals. The more I learned about this practice, the more shocked and disgusted I became. I began to buy only non-animal tested products. Soon, I realized that I could do even more to help the animals by encouraging other people to use non-animal tested products as well, and that I could do this on a large scale through modeling. Finally, my direction was clear!

I founded a revolutionary group called Models with Conscience. Its mission is to radiantly represent cruelty-free products. Models with Conscience operates on the premise that models are irresistibly radiant when their conscience is clear—when they feel completely positive about the product they are presenting, the way they are presenting it, and the fact that no living creature was tortured or killed in the making of that product. Such cruelty-free products include cosmetics, fragrances, and hair care products that are not tested on animals, and clothing and accessories not made with animal fur. To further promote the welfare of animals in our world, Models with Conscience donates a portion of its profits to animal-friendly charities. Models with Conscience reserves the right to refuse service in any situation that is contrary to the model's personal ethics.

The word "ethics" has different interpretations for different people. Yet, to distinguish ourselves from other groups, it is necessary that members of Models with Conscience share the principle of representing cruelty-free products. From this common foundation, each member has the freedom to further personalize his or her own principles, which are respected within the group.

I chose the mountain lion to be the "mascot" of Models with Conscience. In many Native American cultures, animals are associated with specific traits or are seen as conveyers of special messages. Having some Native American heritage myself, I was especially interested in

Putting my feet up in The Red Victorian's Flower Child Room. Skirt and blouse by Esprit. Photo by Russell Tanoue, make-up by Jessica Campo, hair by Danica Winters.

learning about these associations. A tool that helped me do this was *Medicine Cards: The Discovery of Power Through the Ways of Animals*, written by Jamie Sams and David Carson. According to *Medicine Cards*, the mountain lion reminds us to lead through our own example, and to courageously speak and live what we feel is the truth—without demanding that others follow.

When I began Models with Conscience, I felt that the mountain lion's message summarized the essence of Models with Conscience. I still do. By promoting humane products, our members are living examples of how to be fully radiant and successful while respecting animals. Although we are entirely dedicated to this cause, we are not out to convert every single person on the planet. We do our best to respect each individual's unique growth path and understand that it takes time for people to update their belief systems and change their behavior. Meanwhile, we are doing what we feel is right and offering an example that others may follow if and when they choose to.

We have an extensive questionnaire for prospective members in order to discover their opinions on a variety of animal-related issues. Although, at this point, our primary criterion for representing products is that they be non-animal tested and fur-free, I still believe that it is important to know where applicants stand on related topics, to decrease the chance of misunderstandings if we decide to further specialize in the future.

Models with Conscience members relaxing in The Red Victorian's Peacock Suite. Upper row, Left to Right: Julie, Daniel, Amanda, Petal, Jacqueline, Suzanne; lower row: Crystal, myself and Gary. Petal's blouse and skirt, my blouse and faux leather pants by Esprit. Photo by Russell Tanoue, make-up by Jessica Campo, hair by Danica Winters.

There are three main criteria for becoming a Models with Conscience member: a pleasing physical appearance, genuine compassion for animals, and the ability to radiate confidently in front of the camera. It's a rare combination of qualities, but I believe it exists in the men and women whom I have been privileged to work with as a fellow member of Models with Conscience. In the following pages, you'll meet some of these extraordinary individuals.

2

Models with Conscience

> *We were born to make manifest the glory of God that is within us. It is not just in some of us; it's in everyone.* —Nelson Mandela

THIS CHAPTER WILL INTRODUCE YOU TO A few of the exemplary men and women who comprise Models with Conscience. You will meet them through photos revealing their outer beauty and personal essays revealing their inner beauty. Further in the book, you will meet additional members through their photos as well. These members are quite diverse in terms of their background, age, weight, height, nationality, and personal style—showing that kindness can be found in all generations, shapes, sizes, and walks of life.

You may notice that the members profiled here happen to all be white. This is simply because when the book went to press, only Caucasian models had expressed interest in joining Models with Conscience and had passed the evaluation process. However, I believe that beauty and compassion transcend race, and I enthusiastically invite caring people of all races and ethnicities to apply to join the Models with Conscience team. Now, let's meet a few beautiful individuals who are not only up-and-coming models, but role models, proving that humane living can be stylish, fun, and gorgeous! We'll start with Amanda Sehic.

* * *

Amanda's Story

I am an only child and was raised by my grandparents until I was ten. When I was in second grade, the fire department burned down an old abandoned house across the street from us, and left three cats that were living there homeless. At first they were afraid of humans, and we could not even touch them. My grandmother and I fed them.

Nationality: American
Year of birth: 1971
Hair color: dark blonde
Eye color: green
Weight: 125 lbs.
Height: 5 feet 4.5 inches

Charismatic and caring—Amanda in The Red Victorian's art gallery. Blouse and pants by Esprit. Photo by Russell Tanoue, make-up by Jessica Campo, hair by Danica Winters.

Gradually they began to trust us, and eventually became normal house pets. Soon they started having kittens and more kittens, until we ended up with twenty cats.

One day, I saw the kids from my school bus throwing rocks at a scrawny, starving kitten. I picked him up and brought him home. My grandmother said that we could not feed another cat, but tearfully I told her that I could not abandon him, and eventually she let me keep him. Every night he slept curled up on my pillow. We shared a special bond that I will never forget.

Living with those cats taught me that animals have feelings and that it can be just as rewarding to love and be loved by them as it is to love and be loved by another human being. I do not believe that we have any right to torture and kill animals. How is that different from someone more powerful torturing and killing us?

The world is a giant web woven together by each and every living creature. Anything that happens to one strand of life affects the whole web, our world. As the dominant species we carry the most responsibility for that world and for our actions. Working with Models with Conscience is more than just a fun thing to do; I consider it my responsibility as a human being to do whatever I can to promote cruelty-free products.

* * *

Jacqueline's Thoughts

> . . . *Then God said, 'Let us make man in our image, after our likeness; and let them have dominion over the fish of the sea, and over the birds of the air and over the cattle, and over the earth, and over every creeping thing that creeps upon the earth. . .'* —Genesis 1:26

"Dominion" can be defined as "sovereignty," "to have power over," or, "complete and absolute ownership of land." But these words, "sovereignty," "ownership," and "dominion," have, I

Nationality: American
Year of birth: 1931
Hair color: gray
Eye color: green/gray
Weight: 112 lbs.
Height: 5 feet 3.5 inches

believe, a weighty component of responsibility and stewardship. And those in "power," whether small or great, are called upon to preserve and make flourish all things, including the fish of the sea, the birds of the air, the cattle, all the Earth, and everything that creeps upon the earth. To do less is to risk great harm to one's soul, if not to one's body.

On a less cosmic note: As I write, a little black and tan terrier sleeps at my feet. Some perturbation in the dog disturbs her. She stretches and looks up at me. And I know I love her.

* * *

Vibrant and venerable—Jacqueline "shaking hands" with a German shepherd at the San Francisco SPCA. Photo by Rob Robb, make-up by Jessica Campo, hair by Danica Winters.

Handsome and humane—Daniel at The Marine Mammal Center, wearing his own non-leather jacket. Photo by Russell Tanoue, styling by Jessica Campo and Danica Winters.

Daniel's Comments

The most macho thing a man can do is be kind.
—Linda McCartney

Nationality: English
Year of birth: 1973
Hair color: reddish brown
Eye color: blue/gray
Weight: 189 lbs.
Height: 6 feet 2 inches

After boarding school and college, I met my missing part and light of my life Petal, also a member of Models with Conscience. Together Petal and I work as a team trying to put the message across about our passion for marine cetacea, i.e. dolphins and whales. We both work voluntarily for an organization called International Dolphin Watch (IDW). How can a man be kind to animals but still remain masculine? In today's society there is a tendency for men to see being masculine as being aggressive on many levels, whether it's at work or play. However, where it is true to say that being aggressive is part of being a man, it does not imply aggression in a malevolent, violent manner.

A perfect example to help describe what I feel would be to look at the attitudes of ancient tribal cultures. The fundamental part of being a man in these cultures was protection of not only the tribe, but also the animals and environment that sustained it. Sometimes this might have required aggressive actions, but these peoples also understood the balance of what it meant to care for and respect their animal brothers and sisters as equals in the great chain of existence.

Models with Conscience   29

Striking and sensitive—Petal near a visual poem by Sami Sunchild which reads, "Be somebody magnificent—you can do it!" Sweater with faux fur collar by Esprit. Photo by Russell Tanoue, make-up by Jessica Campo, hair by Danica Winters.

They knew, as we today increasingly are beginning to know, that we are all, after all, children of Gaia!

It is my opinion that by caring for the environment and its creatures, I have in no way compromised my masculinity, but have in fact only strengthened myself by fighting for the only thing that truly matters: love for all life!

Through my work and activism, I hope to be able to get the message into other people's heads about respecting our animal brothers and sisters!

* * *

Petal's Comments

I studied fashion at England's Plymouth College of Art and Design, which later led me to do some modeling. But it never felt right and I decided it was not for me. In my heart, I wanted to help others, bring happiness, and live the life I was meant to live on this planet.

Nationality: Italian
Year of birth: 1968
Hair color: black
Eye color: green/olive
Weight 113 lbs.
Height: 5 feet 7 inches

I love all animals and have a special affinity for dolphins, who have introduced me to the wonders of the animal kingdom. I joined International Dolphin Watch (IDW) in June 1996, and later became the Projects Officer, writing to many people telling them of the work IDW did. It was my intention to bring about the awareness of dolphins and dolphin healing, not forgetting their ability to communicate with us. It made me incredibly unhappy to discover that dolphins were being killed and maimed each year through the tuna fishing industry.

Dolphins had such a strong influence in my life that I knew something had to be done. In September 1997, Daniel and I held a very large and successful party to benefit IDW. Through my work with the dolphins, I met Heather and was happy to join Models with Conscience. What could be more apt than working with an organization that has the same motivation and beliefs as I do?

I know that animals are here for a reason, not simply to be looked at and patted on the head. I feel so safe when I am around animals. They are here after all to teach us something about life, to take us up above the mundane in our lives and help us reach new heights.

One truth the dolphins have to teach us is that we don't need to compete, that life is to be shared. I hope to be able to walk my talk and smile a big smile when putting a message across to the world: that we should not be harming the universe and all its creatures!

* * *

Suzanne's Comments

I love animals. My mother shared her love of animals with me when I was a child. I believe animals are special creatures. Although their world is different from the human world, they are soulful beings. Animals offer friendship and companionship that is so wonderful and different from human relationships. Animals teach us about ourselves and about life. They accompany us in the journey of life.

Nationality: American
Year of birth: 1961
Hair color: blonde
Eye color: blue
Weight: 128 lbs.
Height: 5 feet 8 inches

People who care about animals need to be the voice for animals and to ensure they are not treated inhumanely. I feel it is a shared responsibility of all humans to protect animals and try to eliminate all mistreatment of them. It is clear to me that animals have feelings and souls.

Working with Models with Conscience is a great way to put my passion for animals in action. We live in a world where the consumer receives a product and is not aware of the suffering an animal endured to produce the product. People need to be educated about how animals are treated. The abuse and suffering must be stopped, and I hope that my work can contribute to that day when they are.

* * *

Lovely and loving—Suzanne radiates in the spotlight. Photo by Rob Robb, make-up by Jessica Campo, hair by Danica Winters.

Julie and Amanda embracing "Killer" the guinea pig at the San Francisco SPCA. Amanda's tagua nut earrings by One World Projects and sweater with faux fur collar by Esprit. Photo by Russell Tanoue, make-up by Jessica Campo, hair by Danica Winters.

3

Aware Personal Care

> *Vivisection is a social evil because if it advances human knowledge, it does so at the expense of human character.* —George Bernard Shaw

As you rub lotion on your skin, do you ever wonder, "How was this lotion made and what's in it?" Until recently, questions like these rarely entered my mind. But now they do, because I've learned that, often, seemingly harmless personal care products like lotion are made using products that have been painfully tested on animals, or contain ingredients painfully obtained from their bodies. I don't want to contribute to such practices, so now I try to find answers to these questions when I shop, and I only buy products that do not cause suffering to animals.

This chapter will help you do the same. It gives a brief introduction to animal testing and animal-derived ingredients, as well as a convenient list of firms that don't test on animals and a list of substances obtained from animals—tools to make cruelty-free shopping a breeze!

What is Animal Testing?
What is animal testing? Also called vivisection, it is the practice of dissecting, cutting, or harming a living animal, usually in the name of scientific experimentation. Thousands of companies use this cruel practice, causing an estimated fourteen million animals to die each year. If it's hard for you to imagine what fourteen million looks like, think of it as the same number of people who live in New York City, Chicago, San Francisco, Philadelphia, Boston, and Washington, DC *combined*!

One of the most common tests is the Draize Eye Irritancy Test, which is used to determine if a substance may irritate the human eye. The test involves restraining an animal, usually a rabbit, so that only her head protrudes from a cage. Her eyelids are held open with clips while samples of a substance are dropped into her eyes. Symptoms of irritation range from swelling to hemorrhaging of the eye. Usually no anesthesia is given to the rabbit, who often struggles so hard to escape that her neck breaks. The testing period can last up to twenty-one days, after which the rabbits are killed or tested on again.

Another common test is the Lethal Dose Test (LD50), used to determine how much of a substance will kill a percentage of a group of animals—usually mice, rats, or dogs. The test involves forcing caged, unanesthetized animals to ingest a substance through gassing, force-feeding, injection, or application to the skin. This results in symptoms ranging from bleeding, to paralysis, to coma. The testing period can last from two weeks to two years, after which the surviving animals are usually killed.

My affiliated models and I had the good fortune to meet a member of one species often victimized by vivisection during our visit to Maddie's Pet Adoption Center, an exemplary, cageless, no-kill animal shelter run by the San Francisco Society for the Prevention of Cruelty to Animals (SF/SPCA). After visiting the dogs and cats, I asked our guide if the SF/SPCA had any resident mice, rats, or guinea pigs, because I wanted this chapter to feature a photograph of our models with such an animal. She told me the facility didn't have any, and that people hardly ever brought small animals like these into the SF/SPCA.

I was quite disappointed and trying to think of an alternative when, all of a sudden, a man walked in holding...a guinea pig! I was thrilled. I immediately explained what we were doing and asked the man if we could take a few pictures with his friend. "Of course! By all means," he replied enthusiastically.

As we took our photos, the man told us about the adorable critter. His son had nicknamed the guinea pig, "Killer"—quite a misnomer because, in fact, the little animal was completely gentle, trusting, and docile. The man explained that, although "Killer" was not involved in vivisection, he was nearly killed in another way by humans. The man happened to cross a bridge one day and glimpsed another man dangling something off the side. Our man stopped and investigated. As it turns out, a cruel person was holding a helpless guinea pig, and was about to drop it into the water far below. Of course, the little guinea pig would have either been crushed upon impact or would have drowned.

In an act of great compassion, our man bravely rescued the guinea pig and brought him home. The good Samaritan and his son cared for "Killer," but could not keep him, as they already had several other companion animals. So, the man brought him to the SF/SPCA. It was a complete joy for us to hold this precious being. We couldn't imagine why anyone would hurt such a gentle creature in any way.

People protect what they love.
—Jacques Cousteau

Is Vivisection Necessary?

Is vivisection necessary? No law requires that cosmetics, personal care, or household products be tested on animals. Testing a product on animals does not ensure that it is safe on humans. For example, animal testing cannot predict whether a substance may cause an allergic reaction in certain people. Often, potentially harmful products stay on the market because the results of animal tests were inconclusive. In fact, 35,000 people were treated in hospital emergency rooms in one year for cosmetics-related injuries.

Animal testing continues because some companies fear being sued by consumers claiming to have been injured by their products. These companies believe that if they test on animals, they will have a stronger defense in court.

Many alternative testing methods have proven more reliable and less expensive than animal testing. Such methods include performing skin-patch tests on human volunteers, using computer programs to predict the effects of a substance on humans, and testing on human tissues grown in test-tubes (*in vitro*). Whereas animal testing costs an average of $500,000 per product, *in vitro* testing costs only $50,000 per product—one tenth the cost! Companies can also formulate products from ingredients on the long list of substances that are generally recognized as safe.

The most advanced companies have gone a step further, signing the Corporate Standard of Compassion for Animals (CSCA), the highest humane standard set so far. It was created by nine animal protection groups forming the Coalition for Consumer Information on Cosmetics (CCIC). Companies that sign the CSCA promise that, after a set date, they will not conduct or commission animal tests, and will not use any ingredient or formulation that is tested on animals by another company. In short, the CSCA ensures that no animal testing has occurred at any stage of product development by the company, its labs, or suppliers.

The Corporate Standard of Compassion for Animals (CSCA) logo, a trademark of the Coalition for Consumer Information on Cosmetics.

The Coalition has created a logo to print on products conforming to the CSCA (see above). It is designed to help concerned shoppers easily recognize these most humane products.

In the meantime, you can begin shopping humanely by consulting this chapter's list of over 700 reputable companies that test their products for safety without the use of animals. Make sure to take this list with you when you go shopping. Before putting a personal care item in your shopping cart, take a moment to look up the manufacturer on the list. If the company is not there, look for a comparable product made by a firm that is. Then you can buy with confidence.

Fortunately, the list of cruelty-free companies is constantly growing, so I suggest contacting the CCIC or animal protection groups like People for the Ethical Treatment of Animals (PETA) on a biannual basis for the most up-to-date lists. You may also like to contact firms not on the list, encouraging them to adopt non-animal testing methods as well. Businesses exist by virtue of the consumer's demand for their products. So let's demand cruelty-free products!

> The animals of the world exist for their own reasons. They were not made for humans any more than black people were made for whites or women for men.—Alice Walker

Animal-Derived Ingredients
Do you wash your face with cow fat? You are if the soap you use contains an ingredient called tallow. Although a product is not tested on animals, it may still contain substances that

Crystal is vivacious without vivisection, wearing entirely non-animal tested make-up and hair products. Photo by Rob Robb, make-up by Jessica Campo and Russell Tanoue, hair by Danica Winters.

may involve the suffering or death of animals. Many conscious consumers avoid such ingredients, because buying them would support industries that profit from animal exploitation. Some substances are derived only from animals, while others may be animal-derived, plant-derived, or man-made. Usually, the product label does not specify the ingredients' source. Here, I will address substances that are derived only or primarily from animals, and commonly found in personal care products.

Tallow

Let's start with just a few of the many animal-derived ingredients for which there are viable plant-derived alternatives—for instance, tallow. It appears in various forms (tallow acid, tallow glycerides, talloweth-6, etc.) and is obtained from rendered cow fat, a slaughterhouse byproduct. When used in cosmetics and personal care products, it may cause eczema and blackheads. For many applications, an excellent alternative to tallow is vegetable glycerin. Why, then, do some companies still use tallow? Probably because they can buy it from the slaughterhouses cheaply and sell more of their products at a lower price.

Spermaceti

Another example is spermaceti. Even though it may become rancid and cause irritations, spermaceti is used in some personal care products, including skin creams and shampoos.

It is a waxy oil obtained from dolphins or from the heads of sperm whales. Can you imagine killing a whale and cutting open its head just to get an ingredient for your shampoo? Absurd, isn't it? Especially since there is a plant-derived substance with properties almost identical to spermaceti.

This superb alternative is jojoba oil (pronounced "ho-ho-ba"). It comes from the seeds of the jojoba plant, native to the Southwestern region of North America. Since only the seeds are used to obtain jojoba oil, the dolphins, whales, and even the jojoba plants themselves can continue to live while you enjoy clean, shiny hair!

Estrogen
Next, let's consider estrogen. It occurs naturally in women's bodies and decreases after menopause, at which time many women undergo treatments to supplement their levels of estrogen. In fact, commercial estrogen is the most frequently prescribed drug in the United States. It is also found in birth control pills and some skin creams and lotions.

Commercial estrogen is usually extracted from the ovaries of cows and from the urine of pregnant horses. In the latter case, mares are impregnated through artificial insemination and kept in tiny stalls, often so small that the mares cannot even lie down. A container is tied beneath the mares' tails to collect their urine. The horses receive only minimal amounts of water, so the urine will be concentrated and its estrogen levels high.

If the mares do not produce enough estrogen to satisfy the drug companies, they may be sold to rodeos or slaughterhouses. The same fate also usually befalls the mares' foals, most of whom are fattened and slaughtered at an average age of just six months. This happens to an estimated 50,000 foals every year.

There are many alternatives to using animal-derived estrogen and obtaining it in this inhumane way. For use in lotions, simple plant-derived emollients are considered better skin restoratives than estrogen. For hormone replacement therapy, numerous plant-derived alternatives, including phytoestrogens, are available. Further, the symptoms of menopause can often be treated with diet and herbs. For more information on animal-derived estrogen and its alternatives, contact the Humane Society of the United States or the Animal Protection Institute. If you are approaching or experiencing menopause, I also suggest you consult a doctor or health care professional for advice on finding the best therapy program for you.

Civets

As a final example, I would like to share with you information I recently learned about civet farming. Civet farming is the practice of confining African civets to collect and sell their musk, which is used in some perfumes as a preservative. The musk is a glandular secretion of civets, members of the mongoose family similar in size and appearance to foxes.

Let's appreciate horses, not use them as estrogen machines! Gary with Marcy the horse at Hacienda de los Milagros Animal Sanctuary. T-shirt by EnviroLink. Photo by Misha Saez.

Several thousand civets are used for their musk in Ethiopia alone, where most civet farming occurs.

Methods of obtaining the musk are inhumane and have not advanced in a hundred years. Civets are caught in the wild and forced to live the remainder of their lives in cages measuring three feet by one foot by one foot, too small for the animals to even turn around until they lose weight. An estimated forty percent die within the first three weeks. Farmers collect musk from the survivors by scraping the animals' perineal glands, located at the base of the tail, every nine to fifteen days. This can be so painful and traumatic for the civets that some go into shock, refuse to eat, and die.

The remaining civets are subjected to inhumane conditions including: having to breathe thick smoke from nearby fires believed to increase musk production; having no bedding to protect civets from extremely hot days and cold nights; having food contaminated with flies and maggots; and being infested with army ants, which can fill a captive civet's ears and nostrils, leading to suffocation.

There is no need to use natural civet musk in perfumes, because synthetic alternatives have been perfected and are being used successfully in dozens of designer fragrances. How can you help? Use fragrances that do not contain natural civet musk, such as those on the following list. As always though, first check to make sure the company does not test its products or ingredients on animals, in order to insure you are using a fragrance that is truly cruelty-free!

Tallow, spermaceti, estrogen, and civet musk are just some animal-derived ingredients you can now choose to avoid. Several others are presented in a list at the end of this chapter. Feel free to use the list as a guide when shopping, or simply look for products with labels that say, "vegan," or, "no animal ingredients."

* * *

Cruelty-Free Companies

The following selected companies do not test their products or ingredients on animals. Those marked by a tilde (~) comply with the corporate standard of compassion for animals or its European equivalent, and do not test their products or ingredients on animals nor buy ingredients that are tested on animals by other firms. Those marked with an asterisk (*) manufacture strictly vegan products—made without animal ingredients, such as milk and egg byproducts, slaughterhouse byproducts, sheep lanolin, honey, or beeswax. This list is valid as of May 2001 and is based on data from these organizations: BWC USA, IDA, AND PETA. Some words have been abbreviated, for example: Prod = Products; Co = company; Frag = fragrances; Int'l = International; Cos = Cosmetics, etc. A = Austria, B = Belgium, CH = Switzerland, D = Germany, F= France, GB = Great Britain,

ABBA Products, Inc.*
ABEnterprises
Abercrombie & Fitch (The Ltd)
Abkit, Inc. (CamoCare)
Abra Therapeutics*
Advantage Wonder Cleaner*
Alba Botanica~
Alexandra Avery Purely Natural
Alexandra de Markoff (Parlux Fragrances)
Alicina Cosmetic (B)
Allens Naturally*~
Almay (Revlon)
Aloegen Natural Cosmetics (Levlad)
Aloette Cosmetics
Aloe Up
Aloe Vera of America
Alva Umweltschonende Produckte (D)
Alvin Last
Amazon Premium Products*
Amber Aromatherapy (GB)*
American Formulating &* Manufacturing
American International
American Safety Razor (Personna, Flicker, Bump Fighter)
America's Finest Products Corp.*
Amitée Cosmetics (Advanced Research Labs)
Amoresse Labs
Amway

Aware Personal Care

Amyris (GB)
Ancient Formulas
Andrea International Industries
 (Clear Perfection)
Animal Shield (D)
The Apothecary Shoppe
Appearance Matters (GB)
Aramis (Estée Lauder)
~Arbonne International
Ardell International
Argiletz (F)
Arizona Natural Resources
Aromaland
Aroma Vera
ASDA Stores Ltd. (GB)
Astonish Industries*
Atmosa Brand Aromatherapy*
 Products
Aubrey Organics
Aunt Bee's Skin Care
Aura Cacia
Aura-Soma Products Ltd. (GB)
Auromère Ayurvedic Imports*~
Austrian Moor Products (GB)*
The Australasian College of
 Herbal Studies
Autumn-Harp~
Avalon Organic Botanicals~
Aveda
Avigal Henna*
Avon
Ayurherbal Corp.*
Ayurveda Holistic Center*

Badgequo Ltd. (GB)
Bare Escentuals

Barry M. Cosmetics (GB)
Basically Natural*
Basic Elements Hair Care
 System*
Basis (Beiersdorf)
Bath & Body Works
Bath Island
Baudelaire~
BeautiControl Cosmetics
Beauty Naturally
Beauty Through Herbs (GB)
Beauty Without Cruelty*~
Beauty Works (GB)
Beautycology/Candy Kisses
 (USA)*
Beehive Botanicals
Beiersdorf (Nivea, Eucerin, La
 Prairie)
Bella's Secret Garden
Belle Star
Berol (Sanford)
Better Botanicals~
Beverly Hills Cold Wax
Bio-D Company Ltd. (GB)
BioFilm*
Bioforce GmbH (D)
Biogime International*
Biokosma (Caswell-Massey)
Bio Pac*~
Biorganics (GB)*
Bio-Tec Cosmetics
Biotone
Bobbi Brown (Estée Lauder)
Bo-Chem Co.
Body Bistro~
Body Centre, The (GB)*

Body Encounters
Bodyography
Body Reform (GB)
The Body Shop~
Body Time
Bon Ami/Faultless Starch
Bonne Bell
Börlind of Germany~
Botan Corporation*
Botanics Skin Care
Brocato International*
Bronzo Sensualé*~
Brookside Soap Company*
Bug Off*

Caeran
Calder Valley Soap Company Ltd. (GB)*
California Styles
California SunCare
CamoCare Camomile Skin Care Products (Abkit)
Candy Kisses Natural Lip Balm*~
Carina Supply
Carlson Laboratories
Carma Laboratories
Caurnie Soap Company (GB)*
Caswell-Massey
Celestial Body
Chanel
Chatoyant Pearl Cosmetics
Christian Dior
Christine Valmy
Chuckles (Farmavita USA)
CiCi Cosmetics
Cinema Secrets*

Citius USA*
Citré Shine (Advanced Research Labs)
Clare Maxwell Hudson (GB)
Clarins of Paris
Clear Conscience*~
Clearly Natural Products*~
Clear Vue Products*
Clientele
Clinique Laboratories
Colorations*
Color Me Beautiful
Color My Image~
Colourflair Studies Ltd. (GB)
Columbia Cosmetics Mfg.
Comfrey Vertriebs GmbH (D)
Common Scents
Compassionate Cosmetics
Compassion Matters
Conair (Jheri Redding)
Concept Now Cosmetics (CNC)
Co-op (GB)
Cosmonaturel (F)
Cosmyl
Cot 'N Wash*
Country Comfort
Country Save Corp.*
Countryside Fragrances*
Crabtree & Evelyn
Crème de la Terre
Crown Royale, Ltd.*
CYA Products*

Dallas Manufacturing Co.
DampRid, Inc.
Dead Sea Magik (GB)

Decleor USA
Dena Corp.~
Deodorant Stones of America*
Derma-E
Dermalogica
Dermatologic Cosmetic Labs
Desert Essence
DeSoto (Keystone Consolidated
   Industries)
Diamond Brands
Dickinson Brands, Inc.
Diproma (F)
Dolma Vegan Perfumes (GB)*
Dome Cosmetics (GB)
Donna Karan Beauty Company
   (Estée Lauder)
Dr. A.C. Daniels
Dr. Bronner's Magic Soaps*~
Dr. Goodpet~
Dr. Hauschka Skin Care
Dr. Singha's Natural Thera-
   peutics*

Earth Friendly Products~
Earthly Matters*
Earth Science~
Earth Solutions*
Eberhard Faber (Sanford)
E. Burnham Cosmetics
Ecco Bella Botanicals
Echo Essential Oils (GB)
Eco-Dent~
Eco Design Company
Ecover
Edward & Sons Trading Co.
Elizabeth Grady Face First

Elizabeth Peter (A)
Elizabeth Van Buren
   Aromatherapy*~
English Ideas~
Escential Botanicals (GB)*
Espial International*~
Essential Aromatics*
Essential Oil Company
Essential Products of America*~
Estée Lauder (Clinique, Origins)
Eucerin (Beiersdorf)
European Gold
EuroZen*
Eva Jon Cosmetics
Evans International
Every Body, Ltd. (Mountain
   Ocean)
Face Food Shoppe
Faces by Gustavo
Facets/Crystalline Cosmetics
Faith in Nature (GB)*
Faith Products, Ltd.
Farmavita USA (Chuckles)
Faultless Starch (Bon Ami)
Fernand Aubry
Fleabusters/Rx for Fleas~
Fleur Aromatherapy (GB)*
Flower Essences of Fox
   Mountain*
Focus 21 International
Food Lion (house-brand products
   only)
Forest Essentials
Forever Living Products
Forever New International*~
Fragrance Impressions, Ltd.

Framesi, USA
Frank T. Ross (Nature Clean)*
Freeda Vitamins
Free Spirit Enterprises*
French Transit*~
Frische Kosmetik (D)
Fritz Schmidinger (A)
Frontier Natural Products Co-op*
Fruit of the Earth

Gabriel Cosmetics~
Garden Botanika
Garmon Corp.
Georgette Klinger
Gigi Laboratories
Giovanni Cosmetics*
Glitz & Glam (GB)
Golden Pride/Rawleigh
Goldwell Cosmetics
Green Ban*
Green Things (GB)
Gryphon Development (The Limited)
Gunderson Natural Choices*

Hair Workshops (GB)
Hakawerk W. Schlotz GmbH (D)
Halo Purely for Pets~
Hans-Joaquim Brandl (D)
Hard Candy
Harmonie Verte (F)
Hawaiian Resources*~
The Health Catalog
HealthRite/Montana Naturals~
Healthy Solutions~
Healthy Times*

Helen Lee Skin Care & Cosmetics
Hemp Collective, The (GB)*
Hemp Erotica*~
Henri Bendel (The Limited)
Herbal Products & Development~
The Herb Garden*~
Herbes Savantes (F)
h.e.r.c. Consumer Products*
Hewitt Soap Company
High Energy Hair Products (GB)
Hima Laya Natural Cosmetics (D)
Hobé Laboratories
Hoke2~
Hollytrees (GB)*
Homebody (Perfumoils)
Home Health Products
Home Service Products*~
Honesty Cosmetics (GB)
House of Cheriss
H2O Plus
Huish Detergents
Ida Grae (Nature's Colors Cosmetics)

Il-Makiage
ILONA
i natural cosmetics (Cosmetic Source)
Innovative Formulations*
International Rotex*
International Vitamin Corp.
IQ Products Company
Island Dog Cosmetics~
IV Trail Products*

Jacki's Magic Lotion~
Jacques G. Paltz (F)
James Austin Company
Jane (Estée Lauder)
Jason Natural Cosmetics~
J.C. Garet
Jeanne Rose Aromatherapy
Jennifer Tara Cosmetics~
Jess' Bee Natural Lip Balm~
Jessica McClintock
Jheri Redding (Conair)
Joe Blasco Cosmetics
John Amico Expressive Hair Care
John Paul Mitchell Systems*~
JOICO Laboratories*
Jolen Creme Bleach
J.R. Liggett, Ltd.*~
Jurlique Cosmetics

Katonah Scentral
K.B. Products
Kenic Pet Products
Ken Lange No-Thio Permanent Waves*
Kenra Laboratories
Kent Cosmetics (GB)
Kiehl's Since 1851
Kiss My Face~
Kleen Brite Laboratories
KMS Research~
Kobashi (GB)*
KSA Jojoba*~

LaCrista*
Lady of the Lake*~
La Florina GmbH & Co. KG (D)

Lakon Herbals~
LaNatura*
Lander Co.
L'anza Research International*
La Prairie (Beiersdorf)
L'Artisan Savonnier (F)
Laura Paige Cosmetics (GB)
Lee Pharmaceuticals
Lerutan (F)
Levlad/Nature's Gate~
Liberty Natural Products*
Life Tree Products (Sierra Dawn)*
Lightning Products
Lily of Colorado
Lime-Sol Company (The Works)~
Lise du Castelet (F)
Little Forest Natural Baby Prod.*~
Little Green Shop (GB)*
Liz Claiborne Cosmetics
Lobob Laboratories*~
L'Occitane (F)
Logona USA
Lothian Herbs (GB)*
Lotus Light
Louis Widmer (CH)
Louise Bianco Skin Care~

M.A.C. Cosmetics
Magick Botanicals
The Magic of Aloe
Make Up Int'l Ltd. (GB)*
Mallory Pet Supplies
Manic Panic (Tish & Snooky's)
Marcal Paper Mills*~
Marché Image Corp.
Marie M (F)

Marilyn Miglin Institute
Mary Kay Cosmetics~
Masada*
Mastey de Paris
Maxim Marketing (GB)*
Meadowsweet (A)
Meadow View Garden*
Mehron
Mère Cie~
Merle Norman
Mia Rose Products*~
Michael's Naturopathic Progs~
Michelle Lazar Cosmetics, Inc.
Micro Balanced Products*
Mill Creek
Mira Linder Spa in the City
Montagne Jeunesse, Eco-Factory~
Montana Naturals/HealthRite~
Mother's Little Miracle*
Mountain Ocean (Every Body Ltd.)
Mr. Christal's
Murad*
Muse, Body Mind Spirit~

Nadina's Cremes~
Nala Barry Labs*
Narwhale of High Tor, Ltd.
Natracare*~
Natupur Frisor (D)
Naturade Cosmetics~
Natura Essentials*
Natural (Surrey)
Natural Animal Health Products
Natural Bodycare*
Natural Chemistry
Naturally Yours, Alex*

Natural Research People*
Natural Science*~
Nature Clean (Frank T. Ross)*
Nature de France*~
Nature's Acres
Nature's Best (Natural Research People)*
Nature's Country Pet*
Nature's Gate/Levlad~
Nature's Plus
Nature's Radiance
Naturkosmetik (D)
Naturkosmetik Gabrielle - Schachner (A)
Nectar Beauty Shops (GB)
Nectarine
Neocare Labs*
Neo Soma*
Network Management (GB)
New Age Products*
Neway*
Neways
New Chapter Extracts*
New Vision~
Nexxus
Nikken
Nirvana*~
Nivea (Beiersdorf)
No Common Scents
Nordstrom Cosmetics
Norelco*
Norfolk Lavender (GB)
North Country Glycerine Soap
N/R Laboratories
NuSkin Personal Care
NutriBiotic

Nutri-Cell
Nutri-Metics International

The Ohio Hempery
Oliva*
OPI Products
Orange-Mate*
Oriflame USA
Origins Natural Resources (Estée Lauder)
Orjene Natural Cosmetics
Orlane
Orly International
Otto Basics–Beauty 2 Go!

Oxyfresh Worldwide*~
Pacific Scents*
Pamela Stevens Ltd (GB)
Parlux Fragrances (Perry Ellis, Todd Oldham)*
Pashtanch*
Pathmark Stores, (house-brand products only)
Patricia Allison Natural Beauty~
Paul Mazzotta*
Paul Mitchell*~
Penhaligon's (GB)
Perfect Balance Cosmetics
Perfumer's Guild, The (GB)
Perovit-Etol-Werk (A)
PetGuard
Pets 'N People*
Pharmagel International*
Phytoceane (F)
Phytomer (F)
Pierre Fabre (Physicians Formula)

Pilot Corporation of America*
Planet*
PlantEssence
Poppy Seeds Ltd. (GB)
Prescription Plus
Prescriptives
Prestige Cosmetics
Prestige Fragrances
The Principal Secret
Professional Pet Products
Pro-Tec Pet Health~
Protocol Cosmetics~
Provida Kosmetik (D)
Puig USA
Pulse Products*
Pure & Basic Products*
Pur'air (F)
Pure Fantasy Cosmetics (GB)
Pure Plant (GB)*
Pure Touch Therapeutic Body Care*

Queen Helene~
Quinessence Aromatherapy (GB)

Rachel Perry~
Rainbow Research Corporation
The Rainforest Company
Real Animal Friends*
Recycline~
Rejuvi Skin Care~
Reviva Labs
Revlon (Almay, Jean Naté)
Ringana Bio-Bio (A)
Rivers Run*
Ronson Home Shopping (GB)

Royal Herbal
Royal Labs Natural Cosmetics*
Rusk

Sacred Blends~
Safeway (house-brand products only)
Sainsbury's (GB)
Sanford (Berol, Eberhard Faber)
Sanoll Ziegermilch (A)
Santa Fe Botanical Fragrances*
Santa Fe Soap Company*
Sappo Hill Soap Works*
Sassaby (Jane, Estée Lauder)
Schiff Products
Scruples
Sea Minerals~
Sea-renity
Sebastian International (Wella)
Secret Gardens
Senator USA~
SerVaas Laboratories*
Seventh Generation*~
Shadow Lake*~
The Shahin Soap Co.*
Shaklee Corporation
Shikai (Trans-India Products)
Shirley Price Aromatherapy
Shivani Ayurvedic Cosmetics
Simplers Botanical Co.*~
Simple Wisdom
Simply Soap*~
Sinclair & Valentine
Sirena (Tropical Soap Co.)*
Skin Essentials
Smith & Vandiver

The Soap Opera~
Soapworks~
Sojourner Farms Natural Pet Products
Solgar Vitamin Co.
Sombra Cosmetics
Sonoma Soap Company~
SoRik International
Soya System
Spa Natural Beauty
Staedtler, Ltd.
Stanley Home Products
Steps in Health
Sternof Vital-Kosmetik (A)
Stevens Research Salon Products*
Stila Cosmetics (Esteé Lauder)
St. John's Herb Garden~
Studio Magic
Styx-Naturcosmetic Krautergarten (A)
Sukar (GB)*
Sukesha (Chuckles)
Sumeru*
SunFeather Natural Soap Co.~
Sunrider International
Sunrise Lane Products
Sunshine Natural Products*
Sunshine Products Group*
Supreme Beauty Products Co.
Surrey

Tammy Taylor Nails
Tapir (D)
TaUT by Leonard Engelman
Ted Stone Enterprises
Terra Natura (A)

Aware Personal Care

TerraNova
Terressentials*
Tesco Stores Ltd. (GB)
Thursday Plantation
Tish & Snooky's (Manic Panic)
Tisserand Aromatherapy*~
Togal Werk (D)
Tommy Hilfiger (Esteé Lauder)
Tom's of Maine~

Tony & Tina Vibrational
   Remedies
Tova Corporation
Trader Joe's Company
Travel Mates America
Tressa
TRI Hair Care Products
Trophy Animal Health Care
Tropix Suncare Products*
Tuesday's Girl Ltd. (GB)

Un-petroleum Lip Care~
Upper Canada Soap & Candle
   Makers
Urban Decay~
USA King's Crossing (Total
   Shaving Solution)*
U.S. Sales Service (Crystal
   Orchid)*

Vermont Soapworks~
Vegan Verde (D)
Vegana Naturkraft (D)
Veterinarian's Best*
Victoria's Secret
Virginia Soap, Ltd.
Von Myering by Krystina*~
V'tae Parfum & Body Care~

Wachters' Organic Sea Products
Warm Earth Cosmetics*~
Weleda~
The Wella Corporation
   (Sebastian)
Wellington Laboratories
Whip-It Products*
Wind River Herbs
WiseWays Herbals
Womankind
W. Ulrich (D)
Wysong
Yardley (GB)
Your Body Ltd. (GB)*

Zia Natural Skincare

Civet-Free Fragrances

The following fragrance brands do not contain animal-derived civet musk. Parent companies or representatives are shown in Parenthesis. Plus signs (+) indicate brands not tested on animals. Squiggles (~) indicate animal-testing status uncertain. Civet-Use status information is courtesy of wspa. Animal-testing status information is courtesy of bwc usa ida and peta. Valid 9/99.

Aramis (Estee Lauder) +
Burberrys (Fragrance Factory) ~
Bvulgari (Classic Beauty) ~
Cacharel (Prestige And Collections) ~
Calvin Klein (Unilever) ~
Christian Dior ~
Diesel (Fragrance Factory) ~
Dolce And Gabanna (Aspects Beauty Company) ~
Donna Karen (Estee Lauder) +
Elizabeth Arden (Unilever) ~
Escalda (Kenneth Green) ~
Fendi (Fragrance Factory) ~
Giani Versace (Aspects Beauty Company) ~
Giorgio Armani (Prestige And
Aramis (Estee Lauder) +
Burberrys (Fragrance Factory) ~
Bvulgari (Classic Beauty) ~
Cacharel (Prestige And Collections) ~
Calvin Klein (Unilever) ~
Christian Dior ~
Diesel (Fragrance Factory) ~
Dolce And Gabanna (Aspects Beauty Company) ~
Donna Karen (Estee Lauder) +
Elizabeth Arden (Unilever) ~
Escalda (Kenneth Green) ~
Fendi (Fragrance Factory) ~
Giani Versace (Aspects Beauty Company) ~
Giorgio Armani (Prestige And Collections) ~
Givenchy ~
Guy Laroche (Prestige And Collections) ~
Halston (Selective Brands) ~
Hermes (Kenneth Green) ~
Hugo Boss (Procter And Gamble) ~
Jaguar ~
Jean Paul Gaultier (Kenneth Green) ~
Kenzo (Kenneth Green) ~
Lanvin (Prestige And Collections) ~

L'eau D'issey (Kenneth Green) ~
Liz Claibourne (Kenneth Green)
~
Moschino (Aspects Beauty Company) ~
Paco Raban (Creative Fragrances) ~
Quorum (Creative Fragrances) ~
Ralph Lauren (Prestige And Collections) +
Rochas (Cosmopolitan Cosmetics) ~
Tommy Hilfiger (Estee Lauder) +
Ungaro (Classic Beauty) ~
Yves Saint Lauren (Sanofi Beaute) ~

Liberate yourself and animals—use cruelty-free products! Jacqueline at The Marine Mammal Center. Photo by Russell Tanoue.

Animal-Derived Ingredients

The following substances are obtained from animals. This list is based on information from The National Anti-Vivisection Society and People For the Ethical Treatment of Animals. In some entries animal proteins abbreviated AP and similar substances are listed together.

Acetylated Hydrogenated Lard Glyceride
Acetylated Lanolin
Acetylated Lanolin Alcohol
Acetylated Lanolin Ricinoleate
Acetylated Tallow
Adrenaline
Albumen
Albumin
Ambergris
Amerachol (TM)
Ammonium Hydrolyzed Protein
Amniotic Fluid
AMPD Isoteric Hydrolyzed AP
Amylase
Animal Collagen Amino Acids
Animal fats oils skin or hair
Animal Keratin Amino Acids
Animal Protein Derivative
Animal Tissue Extract —Epiderm Oil R
Arachidonic Acid

Batyl Alcohol
Batyl Isostearate
Beeswax Bee Products Cera Flava
Benzyltrimonium Hydrolyzed AP
Blood
Bone Char, Bone Meal

Brain Extract
Buttermilk
C30-46 Piscine Oil
Calfskin Extract
Cantharides Tincture Spanish Fly
Carmine Cochineal
Carminic Acid Natural Red No. 4
Casein Estrogen
Castor Castoreum (not Castor Oil)
Catharidin
Ceteth-2—Poltethylene (2)
Cetyl Ether
Ceteth-2 -4 -6 -10 -30
Cetyl Alcohol, Cetyl Palmitate
Chitosan
Cholesterol, Cholesterin
Civet
Cochineal
Cod-Liver Oil
Coleth-24
Collagen
Cortisone, Corticosteroid, Hydro-cortisone
Cysteine-L-Form
Cystine (or Cysteine)

Dea-Oleth-10 Phosphate
Desamido Animal Collagen
Desamidocollagen
Dicapryloyl Cystine
Diethylene Tricaseinamide
Diglycerides, Glycerin, Monoglycerides
Dihydrocholesterol
Dihydrocholesterol Octyledecanoate
Dihydrocholeth-15
Dihydrocholeth-30
Dihydrogenated Tallow Benzylmoniumchloride
Dihydrogenated Tallow Methylamine
Dihydrogenated Tallow Phthalate
Dihydroxyethyl Tallow Amine Oxide
Dimethyl Hydrogenated Tallowamine
Dimethyl Tallowamine Dimethyl Stearamine
Disodium Hydrogenated Tallow Glutamate
Disodium Tallamido Mea-Sulfosuccinate
Disodium Tallowaminodipropionate
Ditallowdimonium Chloride
Dried Buttermilk
Dried Egg Yolk
Disodium Tallamido Mea-Sulfosuccinate
Disodium Tallowaminodipropionate
Ditallowdimonium Chloride

Egg, Egg Oil, Egg Powder, Egg Protein
Egg Yolk, Yolk Extract
Elastin
Embryo Extract
Emu Oil
Estradiol
Estradiol Benzoate
Estrace
Estrone
Ethyl Arachidonate
Ethyl Ester of Hydrolyzed AP
Ethyl Morrhuate—Lipineate
Ethylene Dehydrogenated Tallowamide

Fish Glycerides
Fish Oil Fish Liver Oil Fish Scales

Gelatin Hide Glue Isinglass
Glucuronic Acid
Glyceryl Lanolate
Glycogen
Guanine—Pearl Essence

Heptylundecanol
Honey
Human Placental Protein
    Human Umbilical Extract
Hyaluronic Acid
Hydrogenated Animal Glyceride
Hydrogenated Ditallow Amine
Hydrogenated Honey
Hydrogenated Laneth-5 -20 -25 Oleyl Alcohol
Hydrogenated Lanolin
Hydrogenated Lanolin Alcohol
Hydrogenated Lard Glyceride

Hydrogenated Shark-Liver Oil
Hydrogenated Tallow Acid
Hydrogenated Tallow Betaine
Hydrogenated Tallow Glyceride
Hydrolyzed Animal Elastin
Hydrolyzed Animal Keratin
Hydrolyzed AP
Hydrolyzed Casein
Hydrolyzed Elastin
Hydrolyzed Human Placental Protein
Hydrolyzed Keratin
Hydrolyzed Silk
Hydroxylated Lanolin

Insulin
Isobutylated Lanolin
Isopropyl Lanolate Isopropyl Palmitate
Isopropyl Tallowate Isopropyl Lanolate
Isostearic Hydrolyzed AP
Isostearoyl Hydrolyzed AP

Keratin
Keratin Amino Acids

Lactic Acids, Lactic Yeasts
Lactose, Milk Sugar
Laneth-5 through -40
Laneth-9 and -10 Acetate
Lanoinamide DEA
Lanolin, Wool Fat, Wool Wax
Lanolin Acid
Lanolin Alcohols, Sterols, Triterpene & Aliphatic Alcohols
Lanolin Linoleate

Lanolin Oil
Lanolin Ricinoleate
Lanolin Wax
Lanosteral
Lard, Lard Glyceride
Lauroylhydrolyzed AP
Lysine Magnesium
Lanolate Magnesium Tallowate

Mammarian Extract
Marine Oil
Mayonnaise
MEA-Hydrolized AP
Menhaden Oil, Pogy Oi
Milk
Milk Protein
Mink Oil
Minkamidopropyl Diethylamine
Monoglycerides, Glycerides
Mossbunker Oil
Muscle Extract
Musk, Musk Ambrette
Myristoyl Hydrolyzed AP

Neat's-Foot Oil

Oleamidopropyl Dimethylamine Hydro. AP
Oleostearine
Oleoyl Hydrolyzed AP
Oleth-10 -2 -3 -25 -50 -5 or -10
Ocenol
Oleyl Arachidate
Oleyl Imidazoline
Oley Lanolate
Ovarian Extract

PEG-2 Milk Solids

PEG-3, -10 or -15 Tallow Aminopropylamine
PEG-15 Tallow Polyamine
PEG-5 through -100 Lanolin
PEG-5 through -70 Hydrogenated Lanolin
PEG-5 to -20 Lanolate PEG-20 Tallowate
PEG-6, -8 or -20 Sorbitan Beeswax
PEG-8 Hydrogenated Fish Glycerides PEG-28 Glyceryl Tallowate
PEG-40 -75 or -80 Sorbitan Lanolate
PEG-75 Lanolin Oil and Wax
Pentahydrosqualene
Pepsin
Perhydrosqualene
Pigskin Extract
Placenta
Placenta Polypeptides Protein
Placental Enzymes, Extract, Lipids & Protein
Polyglyceryl-2 Lanolin Alcohol Ether
Potassium Caseinate or Tallowate
Potassium Undecylenoyl Hydro. AP
PPG-12 - PPG-50 Lanolin
PPG-2 -5 -10 -20 -30 Lanolin Alcohol Ethers
PPG-30 Lanolin Ether
Pregnenolone Acetate
Pristane
Progesterone
Propolis
Purcelline Oil Syn

Rennet, Rennin
Royal Jelly
Saccharide Hydrolysate
Saccharide Isomerate
Serum Albumin
Serum Proteins
Shark-Liver Oil
Shellac
Shellac Wax
Silk Amino Acids
Silk Powder
Snails (usually crushed)
Sodium / TEA-Lauroyl Hydro. AP
Sodium / TEA-Undecylenoyl Hydro. AP
Sodium Caseinate
Sodium Chondroitin Sulfate
Sodium Coco-Hydrolyzed AP
Sodium Hydrogenated Tallow Glutamate
Sodium Laneth Sulfate
Sodium Methyl Oleoyl Taurate
Sodium n-Mythyl-n-Oleyl Taurtate
Sodium Soya Hydrolyzed AP
Sodium Tallow Sulfate
Sodium Tallowate
Sodium Undecylenate
Soluble (Animal) Collagen
Soya Hydroxyethyl Imidazoline
Spermaceti Sperm Oil
Spleen Extract
Squalene
Stearamide, Stearamine, Stearates
Stearic Acid, Stearic Hydrazide, Stearone
Stearoxytrimethylsilane, Stearoyl

Lactylic Acid
Stearyl Alcohol—Stenol
Stearyl Betaine, Stearyl Imidazoline

Tallow Amide
Tallow Amidopropylamine Oxide
Tallow Amine, Tallow Amine Oxide
Tallow Glycerides
Tallow Hydroxyethal Imidazoline
Tallow Imidazoline
Tallow, Tallow Acid
Tallowmide DEA and MEA
Tallowmidopropyl Hydroxysultaine
Tallowminopropylamine
Tallow Trimonium Chloride—Tallow
Talloweth-6
Tallowmphoacete
Tallowmide DEA and MEA
Tallowmidopropyl Hydroxysultaine
Tallowminopropylamine

Tea-Abietoyl Hydrolyzed AP
Tea-Coco Hydrolyzed AP
Tea-Lauroyl Animal Collagen
Tea-Lauroyl Animal Keratin Amino Acids
Tea-Myristol Hydrolyzed AP
Tea-Undecylenoyl Hydrolyzed AP
Testicular Extract
Threonine
Triethonium Hydrolyzed AP Ethosulfate
Trilaneth-4 Phosphate
Turtle Oil, Sea Turtle Oil
Tyrosine, Glycose Tyrosinase

Urea, Imidazolidinyl Urea, Uric Acid, Carbamide

Whey
Wool Fat, Wool Wax Alcohols

Yogurt

Zinc Hydrolyzed AP

Dress to thrill, not to kill. Suzanne trying on a hat in Gigi's Room at The Red Victorian. Photo by Russell Tanoue, make-up by Jessica Campo, hair by Danica Winters.

4

Compassion in Fashion

> *Killing animals for sport, for pleasure, for adventures, and for hides and furs is a phenomenon which is at once disgusting and distressing. There is no justification in indulging in such acts of brutality.*
> —His Holiness The XIV Dalai Lama of Tibet

HOW DO OUR FASHION CHOICES AFFECT animals? This chapter offers some surprising answers. It briefly discusses the fur trade and animal-derived materials used in apparel, and offers a handy list of top designers who do not use fur, and a list of animal-derived materials. It's all designed to help you choose clothes and accessories reflecting your personal style and values—the ultimate fashion statement!

What is the Fur Trade?

The fur trade is an industry in which an estimated 100 million animals are killed each year to provide fur for people to use as clothing and accessories. Those who buy fur usually consider it a luxurious status symbol, but few realize the suffering it involves.

Most fur garments come from wild animals such as raccoons, lynxes, or coyotes who have been trapped. The most common trap used is the leg hold trap. An unsuspecting animal steps into the trap, which slams shut on the animal's leg. The animal may be caught for weeks until he either dies, is found, or escapes. Until then, the animal struggles so hard that he may chew desperately day and night on the metal trap or even on the caught leg or foot in order to cut it off and escape. Even if the animal escapes, he still faces injury-related illness such as infection, and is more susceptible to attack by predators. Leg hold traps are indiscriminate and can injure or kill "non-target" animals such as endangered species and domestic pets.

Another method of obtaining fur is raising and killing animals, such as minks and foxes, on so-called fur ranches. The animals on fur ranches are forced to live and die in appalling conditions. They are kept in tiny wire cages, which are crowded together in sheds. The caged animals often develop repetitive habits such as pacing and destructive behaviors such as gnawing on themselves and other animals. No law ensures humane methods of killing animals on fur ranches, and the methods used are usually those which are the cheapest and keep the fur intact—such as electrocution, painful gassing, decompression, and poisoning.

Members of Models with Conscience were introduced to members of two fur-bearing species at the SF/SPCA. In addition to numerous cats and dogs, the SF/SPCA also houses these two animals and uses them in humane education programs. First, we met a Rex rabbit named Cocoa, who stayed snuggled in a cloth "nest" because she is quite shy. Considering this, we were impressed by how tolerant and calm she remained as we stroked her soft, brown fur and talked about her excitedly with our guide. We learned that Rex rabbits are domesticated animals who like to eat fresh green leafy vegetables like spinach and kale. Sadly, they are also bred specifically for their fur and, according to our guide, 150 of these docile creatures are killed to make a full-length rabbit coat.

Next, we met Miguel the chinchilla, which was an especially exciting experience for us, as most of us had never seen a chinchilla before. Like Cocoa, Miguel was also rather nervous and liked to stay secure in his "nest." Chinchillas may look a bit funny with their large ears, big eyes, and long whiskers. Our guide told us that chinchillas are nocturnal and these features help them to navigate in the dark. We also learned that they are rodents native to the Andes Mountains of South America, where they have been nearly rendered extinct. Most members of this unusual species are bred for their fur or, as in Miguel's case, as domestic pets. Miguel's fur is light gray and felt so soft that we could barely even feel it. Crystal described the sensation best by saying, "It's like running your hand through charcoal ashes."

But, some people aren't content just to pet chinchillas' extraordinary fur. They want to wear it. We were shocked when our guide told us that in order to make a full-length chinchilla coat, 300 of these unique beings are killed!

> *Where you made a faulty choice before you can now make a better one, and thus escape all pain that what you chose before has brought to you.*
> *—A Course In Miracles*

Faux Fur
We can save animals like Rex rabbits, chinchillas, and others by using fur alternatives in clothing and accessories. Man-made materials such as dense knits, microfleece, and synthetic insulating fibers yield as much warmth as animal fur and have even kept arctic explorers warm in sub-zero temperatures. Luxurious-looking substitutes include velvet, satin nylon, acrylic chenille and synthetic (or faux) fur. Recently, I saw a coat in a boutique that looked and felt so

Suzanne caressing Miguel the chinchilla at the San Francisco SPCA. Photo by Rob Robb, make-up by Jessica Campo, hair by Danica Winters.

much like real fur, I had to check the label to find out if it was real or not. I was very pleased to discover that it was entirely synthetic!

If you own a fur coat now, you may consider donating it to an animal protection organization to sell and use the proceeds to fund anti-fur campaigns. Look out for advertisements depicting fur as glamorous or desirable. Don't support companies that advertise in this way, and don't buy magazines that accept such ads. Even items that are trimmed with fur, like jackets with fur around the hood, play a role in the fur trade and continue the cycle of suffering. So, I suggest you steer away from these, too. Rather, choose garments made with faux fur.

Several top-name fashion designers, including the legendary Oleg Cassini (whom you will see later in this book), do not use real fur in their designs and opt instead for high quality synthetic (faux) fur which looks and feels just like real fur. If you are unsure whether an item is animal or synthetic fur, read the garment's label, ask a salesperson, or call the manufacturer. Also, notice the hair in cosmetic brushes you use, which might contain animal hair, and choose those made with synthetic or plant-fiber bristles.

> *In matters of principle, stand like a rock; in matters of taste, swim with the current.* —Thomas Jefferson

Compassion in Fashion

Crystal glows in faux fur—literally! Coat with faux fur collar by Esprit. Photo by Russell Tanoue, make-up by Jessica Campo, hair by Danica Winters.

Other Animal Products We Wear

Leather, snakeskin, silk, wool, kidskin, ivory. . .what images do these words bring to your mind? If some companies had their way, you would see a sophisticated man strutting confidently in his, "premium calfskin oxfords, suit of the finest Marino wool, adorned with a lustrous silk tie," or a glamorous woman steering her luxury car while wearing "the softest kidskin driving gloves, and a hand-carved ivory bracelet," as her "limited-edition designer snakeskin handbag" reclines in the (leather) seat beside her. Such firms want you to associate these items with wealth, prestige, and glamour, so that you will buy and proudly flaunt them.

Actually, the way these materials are produced is nothing to be proud of. Leather is a by-product of the meat industry and accounts for half the economic value of the animal's non-edible parts. Calfskin is the hide of slaughtered veal calves. The calves' mothers, dairy cows, are also killed for their skin if they don't produce enough milk. Leather is also made from the hides of horses, pigs, goats, and sheep after they are killed for their meat. Furthermore, many endangered species (including lizards, zebras, deer, and kangaroos) are often hunted illegally for their hides. Snakeskin is usually cut off the snakes' bodies while they are still alive and fully conscious. Since they are cold-blooded, the then skinless snakes may suffer for hours or even days before they die. Kidskin is obtained from young goats whom ranchers boil alive to soften the skin. Do you want that kind of violence on your hands, or back?

Thankfully, there are humane, high-quality alternatives. For example, "vegetan" is a lightweight synthetic material that is water-resistant, scuff-resistant, and breathable like leather, but made without the cruelty involved in obtaining leather and the pollution involved in tanning it. Vegetan can be dyed various hues and made to look like new leather, antiqued leather, or even suede. When used for footwear, vegetan uppers are backed with approximately seventy percent cotton, so they are partially biodegradable. Vegetan is also made into fake-leather pants, jackets, belts, wallets and other products available through the award-winning English firm, Vegetarian Shoes.

What about silk and down? You might be surprised to learn that millions of silkworms are boiled or steamed alive to obtain the silk fiber from their cocoons. Some people may think, "They're just worms. They can't feel anything, right?" In fact, it has been confirmed that silkworms *can* feel pain.

Down is obtained by plucking insulating feathers from the delicate skin of live geese. The birds usually go into shock after these sessions, which are often repeated several times during their short lives. Silk and down may be considered natural fibers, but, as you can see, they are produced in extremely unnatural ways.

Likewise with wool, especially Merino wool. It is sheared from sheep bred to have unusually wrinkly skin. In summer the abnormally thick wool often causes the sheep to die from heat exhaustion. If they are shorn in winter, they may die from exposure to the cold. Because the skin is unnaturally wrinkly, it frequently gets infested with maggots. So, some ranchers use a technique called "mulesing," which involves slicing off a large section of skin while the sheep is fully conscious. The sheep usually receives no anesthetic and is left with a huge wound that takes up to five weeks to heal, if it ever does. A resulting scar pulls surrounding skin more tightly together. Ranchers who practice mulesing do so because fewer wrinkles mean fewer places for insects to dwell.

In addition to Merinos, other types of sheep also suffer in the wool industry. They may be forcefully held down or trapped in clamps for shearing, during which they are frequently cut by shearing equipment. Often, the sheep are castrated, their ears punched, their tails cut off, and their teeth ground down to the gum—all usually without anesthesia. They are transported in tightly packed ships or trucks, forced to stand constantly, and to go without food or water for long periods of time. If the sheep do not produce adequate wool, they may be brutally slaughtered.

A few wool companies do treat the sheep more humanely, for example those who sell organic wool items available through The EnviroLink Network's Green Marketplace on-line catalog. These sheep are fed one hundred percent organic feed, are not treated with chemicals like hormones or antibiotics, and are free to roam

Kindness is enticing. Petal and Daniel wearing faux leather jackets by Vegetarian Shoes and tagua nut pendants and earrings (on Petal) by One World Projects. Photo by Yolanda Pelayo.

in large, open pastures. They are sheared in a gentle manner, and if their skin happens to get nicked, the shepherd immediately applies an herbal salve to help it heal. If you must buy wool, please choose a more humane type like this.

> *That many [animals] choose, with their powerful forms (horses, elephants, whales, etc.), to cooperate with humans is at times amazing, especially considering the abuse many humans heap upon them. Their ability to forgive and keep trying to teach and help is sometimes beyond human comprehension.—Penelope Smith*

Ivory

Ivory is commonly carved into jewelry and is obtained from the tusks of elephants, whales, and walruses. So many wild animals have been killed for their ivory that several species, including the majestic African elephant—the world's largest land living animal—are now threatened with extinction. I have seen sickening photos showing piles of elephant carcasses with their faces sawed off by ivory hunters. How many of those awe-inspiring animals were parents of young ones, who will now also die without their guardians to feed, teach and protect them?

According to the World Wildlife Fund, many thousands of African elephants die each year as a result of ivory hunting, because trafficking in ivory and other animal parts is big business. The WWF reports that such parts are second only to illegal drugs as the items most frequently smuggled into the United States.

Fortunately, there are cruelty-free substitutes. A material with properties very similar to ivory is the tagua nut (pronounced "tah-gwah"), which grows on a certain type of palm tree in South America's tropical rain forests. Harvesting tagua nuts provides a sustainable industry for the local people, an economic incentive for preserving the rain forests, and is a far more humane alternative to ivory. The tagua nut material is smooth to the touch and can be dyed a variety of colors. Tagua nut jewelry comes in styles ranging

This whole, carved tagua nut by One World Projects reminds us of the magnificent elephants we help protect by choosing jewelry made of tagua nut instead of ivory. Photo by Misha Saez.

from rustic to refined, and is available through a company called One World Projects (listed in the Resources section).

In addition to those discussed above, several other materials are obtained from animals. Some of them are included on this chapter's list. Please use it as a guide to help you avoid buying apparel that was once part of living, breathing beings. Choose cruelty-free items that you can truly wear with pride!

Animal-Derived Materials

The following selected materials often used in beauty or fashions are obtained from animals. Materials of similar origin are listed together.

Alligator hide, crocodile hide
Angora
Animal hair, bone, skin
Animal teeth, claws
Boar bristles
Buckskin
Catgut
Down feathers
Ivory, horn
Kidskin, kid leather
Leather, suede, calfskin

Luna sponge, sea sponge
Pearls, shells
Rawhide
Sable or horsehair brushes
Shatoosh cashmere
Sheepskin
Silk
Snake skin, lizard skin
Tortoise shell
Wool

Compassion in Fashion

Suzanne and Julie laughing over refreshments in The Red Victorian's Friends Room. Photo by Russell Tanoue, make-up by Jessica Campo, hair by Danica Winters.

5

Ethical Eating

> *Nothing will benefit human health and increase the chances for survival of life on Earth as much as the evolution to a vegetarian diet.*
> —Albert Einstein

WE'VE DISCUSSED WHAT WE PUT ON OUR bodies (cosmetics, clothes, etc.), now let's talk about what we put into them (food). It is becoming increasingly evident that eating less—or no—meat is healthier for us, for the animals, and for the environment. The book, *May All Be Fed: Diet for a New*

World by John Robbins, highlights the many benefits of vegetarianism.

For example, people who consume less meat generally enjoy health characteristics like lower cholesterol levels and lower risk for heart attack and some types of cancer. The production of most meat involves such inhumane practices as severe over-crowding, forcing female animals to be continually pregnant (through breeding or artificial insemination), altering the animals' bodies (through hormone injections or genetic engineering) to make them unnaturally heavy, and subjecting them to brutal slaughter methods.

I have lived in areas with factory farms and slaughterhouses. One memory in particular remains with me. I was walking down the sidewalk one day when a truck loaded with pigs slowly passed by in the road just a few feet in front of me. The pigs were packed in the trailer so tightly that everywhere their skin bulged through holes in the metal walls. Their desperate squealing was so loud that I couldn't hear anything else. I sensed their awareness of being taken to slaughter, and their terror. At that moment, I vowed never to eat pork again.

Since then, I have learned more and extend that vow to other meats as well. Recently, I saw a thought-provoking advertisement by the Coalition for Non-Violent Food showing a close-up photo of a cat and a pig, sniffing each other's noses. The caption read, "Which do you pet and which do you eat?. . .Why?" We can each grow a great amount by pondering that one simple question. We allow certain species to share our homes, enrich our lives with their intelligence and love, and call them our "best friends." Meanwhile, though, we confine other species to virtual prison cells, consider them devoid of intelligence and feeling, and call them our "dinner."

We North Americans are shocked and disgusted to hear reports of other cultures eating dogs or horses, because these animals are a few of the species we open our hearts to and wouldn't imagine eating. However, our society is guilty of slaughtering animals that are held dear in other countries. For example, Indians are probably shocked and disgusted at our culture's heartless treatment of cows, which are considered highly sacred animals to Hindus in India. If we have experienced the intelligence and emotional capacity of certain creatures, such as dogs or cats, isn't it possible that other animals also possess these traits? There have been several accounts of "farm" animals performing amazing feats of compassion, such as the pig who rescued a family from a burning house. This and many other accounts are recorded in the books *Peaceful Kingdom: Random Acts of Kindness by Animals*, written by Stephanie Laland, and *Animals as Teachers and Healers: True Stories and Reflections*, by Susan Chernak McElroy. Let's allow these stories to inspire us to give all animals a chance at being more than our food, and let's give ourselves a chance to create a larger circle of "best friends."

From an environmental perspective, animal agriculture is a leading cause of global deforestation, as forests are cut, bulldozed, and burned to create pastures, demolishing the homeland of indigenous peoples and the habitat of native plants and animals. The book, *May All Be Fed*, cites the example of Brazil, where hundreds of millions of acres of rain forest have been destroyed by multinational corporations to produce beef. This process is, author John Robbins reveals, "replacing the world's oldest and richest ecosystems, home to two million or more species of plant and animal life, with a single crop—pasture grass for cattle." As you can see, even though the final product may have an elegant name like *fillet mignon*, extremely inelegant things took place behind the scenes to produce it. Choosing plant-based foods helps put an end to these practices and benefits our bodies, the animals, and the environment.

Just as with other meats, eating fish has negative aspects. These days, fish live in waters polluted by pesticides and other chemicals, which become more concentrated as smaller organisms are eaten by larger ones. According to *May All Be Fed*, since fish are at the top of a long food chain, they are often carriers of hazardously high levels of insecticides, pesticides, and other toxic substances. Such chemicals are then passed on to people who eat the fish.

In addition to fish, various other species of sea life are also affected by the fishing industry. Dolphins, sea turtles, sea lions, and seals may be caught accidentally by the huge nets used by commercial fishers. The nets, called factory trawlers, are the size of football fields and catch tens of thousands of fish at a time. Usually, commercial fishers keep the most profitable fish and dump the others (half a billion pounds of dead and dying fish per year, People for the Ethical Treatment of Animals reports) back into the water.

If they are not caught in nets, dolphins, sea turtles, sea lions, seals, and even birds may still be harmed by fishing. Fishermen often view marine animals who eat fish as a threat and may kill these creatures to preserve their catch. I discovered this firsthand during a photo session at The Marine Mammal Center (TMMC), a Northern California clinic, which rescues, rehabilitates, and releases injured wild ocean mammals. There, my fellow models and I noticed one especially active sea lion with a large, rectangular black patch on his neck.

We asked our guide about the patch and learned that it was a skin graft attached by TMMC surgeons to treat a gunshot wound. I couldn't imagine why anyone would shoot a harmless sea lion, so I asked our guide. She told me that, if this sea lion's case was like many others TMMC had seen, he had probably been shot by a fisherman. Again, I asked, "Why?!" Our guide replied, "Because sea lions eat fish."

Later, we decided to take a photo by this sea lion's temporary enclosure. To our delight, as we knelt beside the fence, he jumped out of his pool, lumbered down a ramp, and posed directly behind us!

Let's cherish all animals—horses, dogs, and cows alike! Petal hugging Josephine the horse at Hacienda de los Milagros Animal Sanctuary. Faux leather jacket and shoes by Vegetarian Shoes. Photo by Yolanda Pelayo

Jacqueline at The Marine Mammal Center with two California sea lions, a species sometimes harmed by the fishing industry. Photo by Russell Tanoue.

> *There is something we can do. . . and it may be the most effective thing we can do for animals in our lifetimes. Go Vegetarian.*
> —Ingrid E. Newkirk

Becoming Vegetarian and Vegan

The strictest vegetarians, called vegans, exclude all animal foods (including eggs, dairy products, even honey) from their diets. You may wonder, "Why say no to these products? After all, the animals aren't killed or hurt to get them, right?" Actually, they can be.

Let's take the example of dairy products. Much has changed since the days when dairy cows roamed freely on large, green pastures, became pregnant naturally, and were milked by hand. According to People for the Ethical Treatment of Animals, most dairy cows now live on factory farms in crowded, concrete-floored pens. They are kept pregnant at all times through artificial insemination on so-called "rape racks," and are milked two to three times each day by machines that may frequently cut the cows' udders and cause infection. Calves are taken away from their mothers only a day or two after birth. The females are usually used as dairy cows or killed for the rennet in their stomachs, which is used to make cheese. Male calves are often put into crates so small the calf cannot turn around, deprived of important nutrients, and forced to live in their own waste for twelve to sixteen weeks until they are killed and sold as veal.

What's more, factory farmers often spray cows with pesticides and/or inject them with tranquilizers, antibiotics, and hormones, some of which can cause the cows' udders to swell so large they may drag on the ground. The cows may inadvertently step on their unnaturally heavy udders, resulting in cuts and infections. Puss from these infections can run into the milk and, along with residues of the hormones and other drugs, be transferred to people who consume the milk or items made from it.

To avoid this, you may choose dairy products from cows that are not treated with these chemicals, and are living in humane conditions. One company offering such items (including milk, yogurt, butter, sour cream, rennetless cheese, etc.) is Horizon Organic Dairy. Its cows are not treated with hormones, pesticides, or antibiotics, are fed one hundred percent organically grown hay and grains, and have access to fresh air, clean water, and exercise.

Perhaps you'd like to go one step further and give up dairy all together. This may well be a healthier choice. Dr. Neal Barnard, President of the Physicians Committee for Responsible Medicine, reports that eating dairy products can have negative effects on human health, including risk for heart disease, diabetes, cancer, and osteoporosis (high levels of protein in dairy and other animal products can actually leach calcium from the body). The United States' recommended daily amount of calcium may be obtained instead, according to Dr. Barnard, from plant foods like broccoli, beans, kale, tofu (soy bean curd), dried fruits, seeds, and nuts.

It's easier than ever to find convenient and delicious meat and dairy alternatives. Here are some alternatives I enjoy. At breakfast, I might have a bowl of hot or cold cereal topped with smooth "milk" made from rice, oats, almonds, or soy beans, such as vanilla-flavored Edensoy®. It is produced by Eden Foods, Inc., using organic soy beans. When lunchtime rolls around, I may enjoy a meatless Gardenburger. Produced by Gardenburger®, Inc., these patties are available in a wide variety of styles including "Hamburger" style, which is very similar to the animal-based variety in taste and texture. For dinner, I often replace ground beef with fat-free Veggie Ground Round, made by Yves Veggie Cuisine®, Inc., and use it in taco filling, Sloppy Joes, spaghetti sauce, etc., accompanied by a fresh salad. As a grand finale, I may savor a scoop of refreshing fruit sorbet or an ice cream substitute such as Turtle Mountain's organic Soy Delicious (my favorite flavor is mint marble fudge).

To add extra nutrients to my diet, I also take herbal, vitamin, or mineral supplements. Even these can be vegetarian if enclosed (not in gelatin, but rather) in vegetable-based capsules like Vegicaps, by Vegicaps Technologies, USA (which, by the way, were developed by Amanda's step-dad!). You will find these and a vast array of other convenient and luscious vegetarian and vegan items at virtually all natural foods stores and select supermarkets.

Some stores offer free samples, so that you can test products before buying them. Vegetarian eating can be a fun adventure for your taste buds, offering you better health, a clearer conscience, and thus, a more vibrant you. How about starting with the scrumptious foods on the following list? *Bon appetit!*

Vegetarian dining can be paradise! Pasta with fresh tomato and basil sauce, served on a veranda overlooking the turquoise Caribbean Sea. Photo by Heather Chase.

Humane Food Alternatives

The following food alternatives are the most similar to animal-based foods that I have found in my years of vegetarian eating. All the suggested brands are made with vegetarian ingredients and some are vegan. They are available at nearly all natural foods stores and many mainstream grocery stores throughout the U.S. Most of the brand and manufacturer names are registered trademarks. The first line of each entry is the animal food. The second line of each entry is the non-animal substitute, and the last line(s) is the producer of the non-animal product.

Ground beef
textured vegetable protein
Veggie Ground Round/
Yves Veggie Cuisine®

Hamburger patties
hamburger style veggie patties
Gardenburger® flame grilled hamburger style/Gardenburger® Inc.

Steak
grilled portabella mushroom caps
(Recipes available in several vegetarian cook books)

Meat hot dogs
veggie dogs
Smart Dogs!®/Lightlife Foods Inc.

Bacon strips
veggie bacon
Morningstar Farms® Breakfast Strips/Worthington Foods Inc.

Bacon bits
veggie bacon bits
Betty Crocker® Bacos®/
General Mills Sales Inc.

Meat corn dogs
veggie corn dogs
Morningstar Farms® meat-free Corn Dogs/Worthington Foods Inc.

Sausage links
meatless sausage links
Morningstar Farms® Breakfast Links/Worthington Foods Inc.

Sausage patties
meatless sausage patties
Morningstar Farms® Breakfast Patties/Worthington Foods Inc.

Meat cold cuts
meatless cold cuts
Yves Veggie Slices/
Yves Veggie Cuisine®

Pepperoni
meatless pepperoni
Veggie Pizza Pepperoni/
Yves Veggie Cuisine®

Salmon
faux salmon
Vegetarian Fillet/
Veat™ Gourmet Inc.

Liver pate
olive and garlic spread
Moshe & Ali's World Famous
Olive Sprate®/PeaceWorks™

Turkey
meatless turkey
Tofurkey/Turtle Island Foods Inc.

Chicken patties
chicken style veggie patties
Morningstar Farms® Chick
Patties®/Worthington Foods Inc.

Chicken nuggets
chicken style veggie nuggets
Chicken-free Nuggets/
Health is Wealth® Products Inc.

Chicken chunks
textured vegetable protein chunks
Vegetarian Gourmet-Bites/
Veat™ Gourmet Inc.

Buffalo wings
veggie drumettes
Morningstar Farms® meat-free
Buffalo Wings/
Worthington Foods Inc

Eggs
egg replacer or cage-free eggs
Humane Harvest™ cage-free
eggs/Chino Valley Ranchers

Milk
non-dairy beverage
Edensoy®/Eden Foods Inc.

Cream
non-dairy creamer
Silk™ Soymilk Creamer/
White Wave Inc.

Whipped cream
non-dairy topping
Hip Whip™/Now & Zen Inc.

Butter
margarine
Smart Balance® Buttery
Spread/GFA Brands Inc.

Yogurt
non-dairy yogurt
Silk® Cultured Soy/
White Wave Inc.

Cheese
non-dairy cheese
Rice™ Slice/Soyco Foods®

Cream cheese
non-dairy spread
Better Than Cream
Cheese®/Tofutti® Brands

Ice cream
non-dairy dessert
Soy Delicious™ non-dairy frozen dessert/Turtle Mountain Inc.

Milk chocolate
non-dairy chocolate
Sweet Dark dark chocolate/
Newman's Own® Organics

Pudding
non-dairy pudding
SuperPudding™/
The Hain Celestial Group Inc.
(make with soy milk)

Gelatin dessert
agar or carrageenan-based dessert
SuperFruits™ Dessert Mix/
The Hain Celestial Group Inc.

Gelatin capsules
vegetable-based capsules
Vegicaps®/
Vegicaps® Technologies U.S.A.

Animal protein bars
vegetable protein bars
Boulder Bar® Endurance/Boulder Bar Endurance™ Inc.

Mayonnaise
egg-free mayonnaise
Vegenaise® dressing and sandwich spread/
Follow Your Heart®

Gravy
animal-fat-free gravy
Vegetarian Brown Gravy Mix/
The Hain Celestial Group Inc.

Honey
agave nectar or maple syrup
Pure Maple Syrup/Maple Grove Farms® of Vermont

Ethical Eating

Amanda luxuriating in the Cat's Cradle Room of The Red Victorian, a humane hotel. Pants and faux fur purse by Esprit. Photo by Russell Tanoue, make-up by Jessica Campo, hair by Danica Winters.

6

More Ways to be Cruelty-Free

> *The way you heal the world is you start with your own family.* —Mother Teresa

So far, BEAUTY WITHOUT THE BEASTS HAS shared ways to help us make humane choices in your personal care, apparel, and diet. I hope it has inspired you to look beyond sly corporate marketing schemes and follow your own principles. This chapter offers ideas for making compassionate choices in other areas of your life as well.

Entertainment

Let's start with our entertainment choices. As you read a newspaper or magazine article, or watch a movie, TV show, or commercial do you notice how animals are used or referred to? Such media outlets have great potential to influence the public's attitude about animals. Which outlets promote compassion for animals, and which do not? This question is the focus of The Ark Trust, Inc., the only animal rights organization dedicated primarily to encouraging positive coverage of animal issues by the major media. The Ark Trust produces the annual Genesis Awards, a formal televised event recognizing outstanding media figures who are increasing public awareness of animal issues. I have had the great honor of being the awards plaque presenter at The Genesis Awards, and I can tell you that meeting celebrities with the courage and integrity to use their talents to help animals is immensely inspiring.

In addition to watching The Genesis Awards, you may also write to actors, authors, TV networks, etc. encouraging them to promote kindness to animals through their work.

Travel

Can our travel choices be humane, too? Yes. Numerous hotels and tour companies around the world feature non-animal tested toiletries, vegetarian meals, and few or no animal-derived materials (down comforters, wool rugs, etc.). One such humane hotel is The Red Victorian Bed, Breakfast

More Ways to Be Cruelty-Free

Heather (left) with designer Oleg Cassini as he accepts an award for his collection of top-quality, synthetic "Evolutionary Furs" at the 13th Annual Genesis Awards. Actress Ali Landry (right) is wearing a faux chinchilla coat from Mr. Cassini's line. Photo courtesy of The Ark Trust, Inc.

& Art, an historic San Francisco art gallery and bed and breakfast.

During a photo shoot for this book, Models with Conscience stayed at The Red Victorian, enjoying its vegetarian breakfasts and its dedication to peace between all life forms expressed in art created by the inn's founder, Sami Sunchild. Our group visited other parts of the San Francisco Bay Area via Birch Circle Adventures, which offers fun, animal-friendly outdoor adventures throughout North America. You may find additional animal-friendly hotels and tour companies listed in a superb book called *The Vegetarian Traveler*, by Jed and Susan Civic and published by Larson Publications.

Tourist attractions often involve animals, who are treated poorly and sometimes even killed in vendors' efforts to "entertain" you. Examples include bull-, cock-, and dog-fights, in which animals are forced to fight to the death. To get otherwise gentle creatures to create a dramatic spectacle, rodeo performers often use inhumane methods such as shocking horses and cows with electric prods, yanking relentlessly on their tails, and cinching straps tightly around their abdomens. The animals buck, not so much to throw the rider off, but rather to get away from whatever is causing the harsh pain. Circuses frequently involve behind-the-scenes brutality such as electrocuting or beating animals into submission and transporting them in overcrowded, life-threatening conditions. As a result of these unnatural and stressful situations, circus animals have been known to

Animal-Free Circuses

Circus Chimera*	Make a Circus
Cirque Ingenieux	Mexican National Circus
Circus Oz	The New Pickle Family Circus
Circus Smirkus	* Circus Chimera includes animals
Cirque du Soleil	in its shows but not as performers.
Cirque Eloize	
Earth Circus	

attack their keepers or spectators. If you want to support circuses that are safer for both animals and humans, I suggest attending animal-free circuses such as those on the previous list, based on information from the Animal Protection Institute.

Suffering inherent in other "attractions" may be less obvious. Horse-drawn carriage rides, for instance, may seem quaint to tourists. But they are not quaint for the horses, who end up breathing noxious automobile exhaust while pulling heavy loads, being deprived of food and water for hours on end, and facing the risk of collapsing from exhaustion or being hit by cars. Realizing the unkindness of such events, several cities and countries have banned them.

> *The greatness of a nation and its moral progress can be judged by how its animals are treated.*
> —Mohandas Gandhi

Responsible Investments

Our values are also reflected in how we invest and donate our money. Many conscious investors are choosing environmentally and socially responsible investments. In a similar spirit, several brokerages now offer portfolios with high animal protection standards. One package, for example, invests only in companies that do not conduct animal testing, do not sponsor the cruel use of animals in entertainment, or do not in any way cause pain or suffering to animals.

Another portfolio adds to these criteria that the firms must not harm the environment, and that their subsidiaries and parent companies must also adhere to these guidelines. For a list of brokerages offering such programs, consult People for the Ethical Treatment of Animals or the Humane Society of the United States.

Did you know that some health charities use donations to pay for painful experiments on animals? Well-intentioned donors think their funds are helping to alleviate suffering, but little do they know they are actually contributing to it. Here is a shocking example of one such experiment. The Physicians Committee for Responsible Medicine reports that a well-known charity paid for research in which the eyes of newborn kittens were sewn shut, the kittens were left like that for a year, and then they were killed, in order to study how not being able to see would affect the cats' brains. All this was done even though it is already an accepted scientific fact that vision changes can affect the brain. Charities like these may have some honorable intentions, such as seeking to treat human diseases, addictions, or injuries. But is torturing and killing animals the best way to reach this goal?

Many kind donors think not. They respect all life by supporting health charities that do not perform animal experiments. There are many to choose from, as scores of US institutes and foundations are working to treat a wide range of afflictions, from AIDS to war injuries, without harming animals. A list of these compassionate organizations is available from the Physicians Committee for Responsible Medicine.

Similarly, when donating to an environmental conservation charity, most people assume their money is going to help protect the environment and all the creatures within it. However, some of these organizations favor certain animal species over others. For example, a group may advocate killing an abundant species, such as deer or feral cats, to protect a rare species, such as endangered plants or birds. Groups with this type of approach often reject more holistic wildlife management techniques, such as humanely trapping members of the abundant species, sterilizing them so they cannot reproduce, and releasing them back into their habitat. If you want your donations used to protect the environment and all of its inhabitants, I suggest researching the conservation groups you are interested in. A list of such groups and their positions on hunting and trapping is available in the book, *Animal Rights: A Beginner's Guide*, by Amy Blount Achor.

> *If a rich man is proud of his wealth, he should not be praised until it is known how he employs it.*
> —Socrates

Crystal and Gary playing with Kona the rottweiler in her "apartment" at the San Francisco SPCA. Gary's tagua nut pendant by One World Projects. Photo by Rob Robb, make-up by Jessica Campo, hair by Danica Winters.

Compassionate Labor

The concept of cruelty-free can be expanded even further, to encompass not only animals, but also people and the Earth. With reference to people, some garments sold in the West are made by exploited workers in sweatshops, usually located in developing countries.

Workers (possibly even children) in these situations frequently endure inadequate wages, unhealthy working conditions, excessive working hours, and abusive treatment from employers. Workers often feel compelled to remain because they have no other employment opportunities and fear that, without the meager earnings, their families would starve. Meanwhile, the items are sold at large mark-ups to Western consumers, who have no idea the conditions under which the products were made.

Many programs are working to end this cycle. In 1997, then-US President Bill Clinton appointed The Apparel Industry Partnership, which created The Workplace Code of Conduct. The Code is a set of standards governing factories in the textile, apparel, and footwear trades. Eventually, the Partnership plans to certify companies as "No Sweat" companies. The US Department of Labor has created the Trendsetters List, an index of companies, including Esprit de Corp., which take exemplary steps to avoid buying apparel made in sweatshop conditions. To learn about these programs, contact the US Department of Labor's Wage and Hour Division.

Peace is not merely a distant goal that we seek, but a means by which we arrive at that goal.
—Dr. Martin Luther King, Jr.

Treating Each Other Nonviolently

Another approach to being cruelty-free toward people is to treat them nonviolently, even when they do things we believe to be wrong. For example, if we hear about a horrible case of animal abuse, naturally we are shocked that the offender could be so heartless. Sometimes, this reaction turns into anger and an impulse for immediate physical retaliation against the abuser. Animal abuse is clearly wrong and calls for serious correction. However, I believe the methods of correction should be legal and nonviolent. After all, if our goal is to reduce the total level of violence in the world, reacting to violent situations with more violence would only add to the problem and negate the goal. Instead, it would be more effectively served through nonviolent means such as peaceful demonstrations, letters of protest, legal prosecution, and/or humane education.

An extraordinary organization, called the Animal Legal Defense Fund (ALDF), works to prosecute animal abusers effectively and promote stronger anti-cruelty laws. ALDF has a roster of over 700 lawyers and law students nationwide who donate their services for this cause. If you see or hear about a case of animal abuse or neglect, contact your local animal shelter, animal control agency, or police department.

Embrace peace for all beings. Crystal with peace symbol painting by Sami Sunchild. Coat with faux fur collar by Esprit. Photo by Russell Tanoue, make-up by Jessica Campo, hair by Danica Winters.

Follow up and ask how they handled the situation. If they refuse to investigate, to remove animals at risk, or to file charges with the local prosecutor, call ALDF's Animal Cruelty Actionline at 800-555-6517.

Prayer
In addition to all of the external actions we can take to protect animals, I add what I believe is a very powerful internal technique, namely prayer. When I hear about animal abuse, I pray for the animals to be freed from their suffering and given the chance to live a decent life. I also pray for the people who are inflicting the pain to be freed from whatever sadness, anger, or insensitivity is causing them to act out in abusive ways, and for their hearts to be opened to the joy and love that can come from respectful interaction with animals.

> For whatever happens to the beasts, soon happens to man. . .Whatever befalls the Earth, befalls the sons of the Earth.—Chief Seattle

Helping the Earth
How can our fashion choices help the Earth? One way is by wearing organic cotton apparel. According to The EnviroLink Network, while conventionally grown cotton uses only three percent of the world's farmland, it uses twenty-five percent of the chemical fertilizers and pesticides.

Besides killing countless insects (including beneficial pollinators), these chemicals can seep into our groundwater and cause serious health problems. EnviroLink reports that forty-eight people *a minute* are poisoned by pesticides worldwide.

These chemicals can harm wild and domestic animals as well. This I know from my own experience of living for a time in a rural area, surrounded by conventional farms. One day my cat, Mikey, went out exploring and came back with no fur on the bridge of his nose! Instead, there was only bare, red skin covered in sores. The veterinarian who examined Mikey determined that the wound was a chemical burn, most likely caused by pesticides. Thankfully, the injury healed and Mikey's fur grew back, but it took several weeks for it to do so.

As I suggested, we can help reduce the use of pesticides by choosing organic cotton. Not only is it grown without harmful pesticides or herbicides, it also a soft, high-quality fiber that is quite pleasant to wear. One source of organic cotton clothing is EnviroLink's on-line catalog, Green Marketplace. Some organic cotton T-shirts offered there have yet another beneficial feature. They are adorned with paintings by Koko, a remarkable western lowland gorilla who communicates through sign language. Koko's paintings are colorful and energetic, and those such as one entitled "Love," show that animals do indeed understand and experience meaningful emotions. A portion of the proceeds from the sale of these unique shirts helps The Gorilla

Foundation, which cares for Koko, to create a seventy-acre sanctuary for Koko and other members of her endangered species. By wearing items like these shirts, we can look *and* feel great, because we are supporting the well-being of ourselves, the animals, and our precious planet.

These are just a few of the myriad ways to be cruelty-free. For more ideas, I encourage you to read *Save the Animals!: 101 Easy Things You Can Do*, by Ingrid Newkirk, and *Animal Rights: A Beginner's Guide*, by Amy Blount Achor. I hope you implement the suggestions in *Beauty without the Beasts* and the two books listed above, then find additional ways to help animals as well. Let your life be a beautiful expression of who you truly are!

Yes, the world has many problems, but there are countless creative, ingenious—even fun—solutions just waiting to be discovered. Each of us really can make a difference and leave the planet better than we found it. Remember the message of mountain lion, and like the members of Models with Conscience, lead through your own example and inspire others to adopt a more humane lifestyle. Use your talents, creativity and intelligence to invent your own unique approaches to helping yourself, other people, the animals and the Earth we all share.

More Ways to Be Cruelty-Free

Julie helps animals and the Earth by wearing an EnviroLink organic cotton T-shirt with "Love" design painted by Koko the gorilla. Julie and Marigold the golden retriever at Hacienda de los Milagros Animal Sanctuary. Photo by Misha Saez.

7

Cosmic Connections

> By ethical conduct toward all creatures, we enter into
> a spiritual relationship with the universe.
> —Albert Schweitzer

OUR BEHAVIOR TOWARD ANIMALS IS AN expression of our beliefs about them. For example, someone who abuses animals most likely sees them as "lower" creatures put on Earth solely to serve humans. Such a belief may have been derived from that person's spiritual tradition. I respect all faiths and honor each individual's right to follow his or her own path. Yet, I wonder if some traditions have been misinterpreted over time, the original intent of their teachings altered through numerous translations or recounts by various followers. We see this phenomenon when a secret is whispered from person to person in a circle. By the time the secret reaches the last person, the words are usually completely different from the words whispered to the first person.

Perhaps we could gain some new insights by researching our faiths, looking especially for their original teachings about animals. I have begun to do this myself and have discovered humane stories from various traditions. Of course, I am not a religious scholar, but I'll share a few of these stories with you as accurately as I can.

Saint Francis of Assisi, Catholicism's patron saint of animals and ecology, was reportedly able to communicate with animals. He is often depicted in sculptures or paintings surrounded by his animal friends, who felt entirely safe being near him. It is said that one time, a village was being terrorized by a man-eating wolf. The desperate villagers did not know what to do and considered killing the wolf. But Saint Francis offered to help. He calmly went into the woods, communicated with the wolf, and discovered that the wolf was eating people simply because he was hungry. Saint Francis proposed an alternative. He asked the wolf if he would leave the people alone if they left food out for him. The wolf agreed. Saint Francis returned to the village and explained the arrangement. The people understood, cooperated, and lived peacefully alongside the wolf ever after. By some accounts, the wolf even protected the village!

Through the organization Jews for Animal Rights, I learned the story of how Moses was chosen by God to lead the people of Israel. Exodus Rabbah 2:2 explains that Moses was tending a flock of sheep when one ran away from him. Moses ran after it and, at last, found it drinking from a pool of water. Instead of being angry at the sheep, Moses said, "I did not know that you were running because [you were] thirsty. You must be tired." He lifted the sheep and carried it on his shoulder. Then God said to Moses, "You are compassionate in leading flocks belonging to mortals; I swear you will similarly shepherd my flock, Israel." Moses was chosen, not because he was an outstanding warrior, speaker, or politician, but rather because he was compassionate toward animals!

The book *Peaceful Kingdom: Random Acts of Kindness by Animals* contains a chapter of compassionate acts humans have done for animals, including one performed by Mohammad, prophet of Islam. The tale recounts a time when a cat had fallen asleep on the wide sleeve of Mohammad's garment. A logical solution would have been to simply pull the sleeve out from under the cat. But, the cat was sleeping soundly and Mohammad did not wish to disturb it. So, the prophet sacrificed his own garment instead. He carefully cut the sleeve off and slid his arm out, allowing the cat to continue sleeping contentedly on the fabric.

A beautiful legend from *Medicine Cards* illustrates the honored role animals often play in Native American tradition. According to this legend, the lodge of Great Spirit was once guarded by a terrible monster who tried to prevent all beings from connecting with Great Spirit. The monster was the very embodiment of all disgusting demons that had ever existed. One day, a gentle fawn encountered the monster and, despite his best attempts to terrify her, the fawn was not frightened at all. She just politely asked him to let her pass, while gazing at him with pure love and compassion in her serene eyes. The fawn's lack of fear shocked and perplexed the brute. Regardless of how the monster tried to horrify her, the fawn continued to emanate love toward him. Soon, the bewildered demon's hardened heart melted and his huge body shrank to the size of a walnut! The fawn's enduring kindness had disarmed and transformed the worst demon imaginable. Thanks to her gentleness and unconditional love, the way is now clear for all beings to connect with Great Spirit.

> *The original instructions of the Creator are universal and valid for all time. The essence of these instructions is compassion for all life and love for all creation.*
> —David Monongye, Native American elder

Daniel walking in harmony with nature while wearing a faux leather jacket and boots by Vegetarian Shoes, an organic cotton T-shirt by EnviroLink, and a tagua nut pendant by One World Projects. Photo by Misha Saez.

Honoring the Animal Voice

From Hinduism comes a tale conveying the value of loyalty between a man and a dog. The man was Yuddhistra, a king who left his kingdom in search of Heaven. The dog was a little stray the king met during his long journey. Together, they walked for several years. Yuddhistra's fellow human travelers each died during the arduous quest, and the little dog remained by the king's side as his only companion. Finally, the two friends reached the Gates of Heaven. A guard said Yuddhistra could enter, but the dog could not. The king was shocked and declared that if his loyal friend could not live in Heaven, then he did not want to live there either. He turned to leave, feeling completely disheartened, thinking the multi-year quest had been useless. Just then, Heaven rejoiced because Yuddhistra had passed his final test of character. The stray dog revealed his true identity—God Himself. The Gates of Heaven opened and God welcomed Yuddhistra into his new, most glorious home.

I have read the above story in several sources, one of which is *Best Friends* Magazine, a publication of Best Friends Animal Sanctuary, North America's largest no-kill animal sanctuary. Best Friends recognizes the spiritual connection between humans and animals in its annual "Conference on Animals and Spirituality." This remarkable sanctuary also occasionally hosts lectures and workshops on telepathic animal communication.

Such communication goes far beyond one-way commands like, "fetch, sit, stay." Instead, telepathic animal communication is a two-way dialogue that can significantly deepen our relationship with, and our perception of, animals. I fully believe it is possible, based on my own experience. When I was around twenty-two I was sleeping early one morning when my cat, Mikey, began scratching on my bedroom door, trying to open it. That didn't work, so he started to meow. Slowly, my awareness transitioned from sleep to wakefulness. I was barely awake, hoping Mikey would be quiet so I could go back to sleep. Then in a clear, sweet, high-pitched voice I heard these words, "Why won't you let me in, please?" Instantly, I was wide awake and my eyes flashed open. I bolted out of bed, opened the door and was greeted by Mikey, who enthusiastically ran straight into my room and jumped onto the bed. I knew it was he who had communicated to me when my mind was quiet and receptive. I hugged him, thanking him for giving me an insight and showing me there was much more to him than just his furry little body.

Since then, I have not had any similar experiences. But I have had sessions with Mikey and my other two cats through professional animal communicators, including former defense attorney Sam Louie, and I have read several books on this fascinating process, such as *Animal Talk: Interspecies Telepathic Communication*, by pioneering communicator Penelope Smith.

Julie, Gary, Petal, Dan, and I experienced animal communication in action during one of the photo shoots for this book. The event was held at Hacienda de los Milagros ("House of the Miracles"—HDLM), a no-kill sanctuary for rescued horses and burros (donkeys). Our photo session was attended by animal communicator Janice Goff, who graciously "translated" between our group and the animals, and conveyed two precious messages from them to us. The first is touchingly funny. Janice told me the female burros liked the color of the leggings I was wearing (taupe), because it made me look like a burro, too!

The second was told by one of the burros, as a representative of her group, to Petal, as a representative of our group. Through Janice, the burro told Petal she wanted to share her breath with Petal. This is how burros greet and welcome each other. Petal leaned over and breathed nose-to-nose with the burro, who then communicated that she and her fellow burros were very happy that we were there. Janice told Petal that such a message was quite unusual, especially since it was the first time our group had met the burros. We all felt truly honored.

Having read and experienced such intelligent, loving messages myself, I am convinced that we humans can each develop the ability to communicate with animals. By doing so, I believe we will gain meaningful insights, come to understand and appreciate animals for the complete beings they truly are, and naturally treat them with the compassion and dignity they deserve.

> *Treat us as you would have others treat yourselves…*
> *Let your heart be touched by our Kingdom.*
> —Sir William, HDLM burro

Animal communicator Sam Louie encourages people to "paws" for reflection, as Julie does here in The Red Victorian's Redwood Forest Room. Skirt by Esprit. Photo by Russell Tanoue, make-up by Jessica Campo, hair by Danica Winters.

Petal embracing Loretta the burro, who shared a profound message with us at Hacienda de los Milagros Animal Sanctuary. T-shirt by EnviroLink with "Bird" design painted by Koko the gorilla. Photo by Misha Saez.

Resources

> *The highest wisdom is loving kindness.*
> —The Talmud

IN CREATING *BEAUTY WITHOUT THE BEASTS*, I obtained much information and assistance from the books and organizations listed below. You may consult them for more details about the topics I have introduced here.

Books
Animal Rights: A Beginner's Guide. Achor, Amy Blount. Yellow Springs, OH: Write Ware Inc., 1996.
>   *Animal Rights* thoroughly explores a wide range of animal rights issues, presents ways we can help, and offers an extensive index of organizations addressing these issues.

Animal Talk: Interspecies Telepathic Communication. Smith, Penelope. Point Reyes Station, CA: Pegasus Publications, 1989.
> *Animal Talk* shares funny and touching anecdotes from the author's "conversations" with a variety of animals, and explains how you, too, can communicate with animals.

Animals as Teachers and Healers: True Stories and Reflections. McElroy, Susan Chernak. New York: Ballantine Publishing Group, 1996.
> *Animals as Teachers and Healers* is a profoundly moving collection of true accounts of animals that transformed and even saved human lives.

May All be Fed: Diet For a New World. Robbins, John. New York: Avon Books, 1992.
> *May All be Fed* explains in depth the benefits of a vegetarian diet for the individual, the animals, the Earth, and society as a whole.

Medicine Cards: The Discovery of Power Through the Ways of Animals. Sams, Jamie and Carson, David. Santa Fe, NM: Bear & Company, 1988.
> *Medicine Cards* is a book and card set explaining the associations Native American cultures attribute to various animals. It can be a powerful tool for learning and growth.

Model: The Complete Guide for Men and Women. Boyd, Marie Anderson. New York: Thunder's Mouth Press, 1996.
> *Model* offers advice for anyone interested in becoming a professional model. It addresses everything from facials to finances.

Peaceful Kingdom: Random Acts of Kindness by Animals. Laland, Stephanie. Berkeley, CA: Conari Press, 1997.

>*Peaceful Kingdom* is a touching and eye-opening collection of true stories describing altruistic deeds performed by animals, both for other animals and for humans.

Save the Animals!: 101 Easy Things You Can Do. Newkirk, Ingrid. New York: Warner Books, 1990.

>*Save the Animals* is a fun book that briefly explains several animal rights issues, and lists simple things each of us can do to help alleviate the problems animals face.

The Vegetarian Traveler. Civic, Jed and Susan. Burdett, New York: Larson Publications, 1997.

>*The Vegetarian Traveler* lists lodgings and tour companies in the US and abroad featuring cruelty-free toiletries, vegetarian menus, and few animal-derived materials.

Zen and the Art of Making a Living. Boldt, Laurence G. New York: Arkana/Penguin Books, 1993.

>*Zen and the Art of Making a Living* contains text, quotes, questions, and exercises to help you find your own unique and fulfilling "life's work."

>*The only person who is educated is the one who has learned how to learn. . .and change.* —Carl Rogers

Jacqueline reading in the Butterfly Room of The Red Victorian. Skirt by Esprit. Photo by Rob Robb, make-up by Jessica Campo, hair by Danica Winters.

Organizations

Animal Protection Institute (API); PO Box 22505; Sacramento, CA 95822; Tel. 916-731-5521; www.api4animals.org

>API is a national non-profit organization dedicated to informing, education, and encouraging the humane treatment of all animals.

The Ark Trust, Inc.; 5551 Balboa Blvd.; Encino, CA 91316; Tel. 818-501-2ARK; www.arktrust.org

>The Ark Trust is a national non-profit group encouraging progressive coverage of animal issues by the major media. The Ark Trust produces the annual Genesis Awards.

Beauty Without Cruelty USA (BWC USA); 175 W. 12th St. #16G; New York, NY 10011-8220; Tel. 212-989-8073

>BWC USA is a tax-exempt charity which exposes the animal suffering involved in the cosmetic and fashion industries, and provides information on cruelty-free alternatives.

Best Friends Animal Sanctuary; Kanab, UT 84741-5001; Tel. 801-644-2001; www.bestfriends.org

>Best Friends is the largest no-kill animal shelter in the USA, offering humane education programs to the public, and housing and care to approximately 1500 animals at a time.

Coalition for Consumer Information on Cosmetics (CCIC); PO Box 75307; Washington, DC 20013; Tel. 888-546-CCIC; www.ddal.org/CCIC

>CCIC is comprised of eight well-known animal protection groups, and offers updated lists of companies having signed the Corporate Standard of Compassion for Animals.

Rekindling the bond—Early humans celebrated their bond with animals through cave art, like that printed on the EnviroLink T-shirts Julie and myself are wearing. Tagua nut jewelry by One World Projects. Photo by Misha Saez.

Coalition for Nonviolent Food, a project of Animal Rights International (ARI); PO Box 214, Planetarium Station; New York, NY 10024
> ARI is a not-for-profit organization that promotes the vegetarian lifestyle and works to reduce the pain and suffering of the more than nine billion animals in factory farms.

Conservation International (CI); 2501 M St. NW Ste. 200; Washington, DC 20037; Tel. 202-429-5660; www.conservation.org
> CI administers numerous environmental preservation programs, including the production and marketing of tagua nuts as a sustainable, cruelty-free alternative to ivory.

Resources

The Gorilla Foundation; PO Box 620-530; Woodside, CA 94062; Tel. 415-851-8505; www.gorilla.org

> The Gorilla Foundation is a non-profit organization running the world's longest inter-species communication project and caring for western lowland gorillas, including Koko.

Humane Society of the United States (HSUS); 2100 L St. NW; Washington, DC 20037; www.hsus.org

> HSUS is an organization working on numerous fronts to create a world in which both animals and humans live harmoniously. Send a SASE for a list of estrogen alternatives.

International Dolphin Watch (IDW); 10 Melton Road; North Ferriby; East Yorkshire HU14 3ET, England; Tel. +44+1482 645789; www.idw.org

> IDW is a non-profit organization founded by Dr. Horace Dobbs to explore the healing effects of interaction with wild dolphins. IDW strives to help both humans and dolphins.

Jews for Animal Rights (JAR)/Micah Publications; 255 Humphrey St.; Marblehead, MA 01945; Tel. 781-631-7601; www.micahbooks.com

> JAR aims to recover the original Jewish instinct that the Earth and all life within it is sacred to God because God created it. Micah Publications is the publishing arm of JAR.

Models with Conscience (MWC); PO Box 1790; Sedona, AZ 86339-1790; www.modelswc.com

> MWC is a revolutionary group of models dedicated to representing cruelty-free products. Enjoy more photos of MWC models and information about us by visiting our website.

National Anti-Vivisection Society (NAVS); 53 W. Jackson Blvd. #1552; Chicago, IL 60604; Tel. 800-888-NAVS (6287), or 312-427-6065; www.navs.org

>NAVS is a national not-for-profit organization dedicated to abolishing the exploitation of animals used in research, education and product testing.

Pegasus Publications; PO Box 1060, Point Reyes, CA 94956; Tel. 415-663-1247; www.animaltalk.net

>Pegasus Publications is *the* resource for Penelope Smith's books, tapes, journal, lectures, journeys, and courses on interspecies telepathic communication.

People for the Ethical Treatment of Animals (PETA); 501 Front St.; Norfolk, VA 23510; Tel. 757-622-PETA (7382); www.peta.org

>PETA is an international nonprofit organization "dedicated to establishing the rights and improving the lives of all animals by educating, changing lifestyles and exposing cruelty."

Physician's Committee for Responsible Medicine (PCRM); 5100 Wisconsin Ave. NW, Ste. 404; Washington, DC 20016; Tel. 202-686-2210; www.pcrm.org

>PCRM, comprised of over 4,000 doctors, works for the health of humans and animals alike, through humane medicine, and provides a list of cruelty-free health charities.

Wage and Hour Division, Employment Standards Agency, US Department of Labor; Washington, DC 20210; www.dol.gov/dol/esa

>This government division compiles the Trendsetters List, and provides information about the Apparel Industry Partnership and the "No Sweat" campaign.

World Society for the Protection of Animals (WSPA); PO Box 190, Boston, MA 02130; Tel. 617-522-7000; www.wspa.org.uk
> WSPA is the world's leading international animal protection charity working to protect, rescue, and heal animals in every corner of the earth.

> *Not to hurt our humble brethren is our first duty to them, but to stop there is not enough. We have a higher mission—to be of service to them wherever they require it.*—Saint Francis of Assisi

Contributors

The following kind-hearted groups and individuals generously contributed goods, services or hospitality to help create this book and support the mission of Models with Conscience. I am profoundly grateful to them, and I enthusiastically invite you to learn more about them.

Animal Legal Defense Fund (ALDF); 127 Fourth St.; Petaluma, CA 94952; Tel. 707-769-7771; ALDF Animal Cruelty Actionline 800-555-6517; www.aldf.org
> ALDF is a non-profit organization with a nationwide network of over 700 attorneys and law students volunteering their services to defend and advance animals' legal rights.

Birch Circle Adventures; 23 Ross Ave.; San Anselmo, CA 94960; Tel. 415-459-7717
> Birch Circle offers fun, animal-friendly outdoor adventures throughout North America, and along with The Red Victorian, offers personal, cultural tours of San Francisco.

Jessica Campo—South Gate Hair Design; Tel. 815-385-2966; Danica Winters—Temptations Hair Design; Tel. 847-854-6800
> Jessica and Danica are licensed cosmetologists specializing in make up artistry and hair design.

Esprit de Corp.; 900 Minnesota St.; San Francisco, CA 94107; Tel. 800-486-4846; www.esprit.com
> Esprit is a socially and environmentally conscious company that designs high-spirited, fur-free clothing, and accessories for women of all ages.

The EnviroLink Network; Tel. 412-420-6400; www.envirolink.org & www.greenmarketplace.com
> EnviroLink offers free Web site hosting to over 400 animal and environmental protection groups, as well as an on-line catalog of animal- and Earth-friendly products.

Janice Goff, Animal Communicator; PO Box 661; Rimrock, AZ 86335; Tel. 520-567-5189
> Janice is a gifted telepathic animal communicator available for consultations and speaking engagements to help humans and animals live together harmoniously.

Hacienda de los Milagros, Inc. (HDLM) Animal Sanctuary; 3731 N. Rd. 1 W.; Chino Valley, AZ 86323; Tel. 520-636-5348; www.enviroweb.org/milagro

> HDLM is a non-profit, no-kill sanctuary for various species, especially rescued horses and wild burros. HDLM utilizes telepathic animal communication with its residents.

Sam Louie, Animal Communicator; PO Box 14741; Berkeley, CA 94712; Tel. 510-644-1583

> Sam is a former attorney who now practices psychic animal communication full-time. He offers consultations and workshops to bring animals and people closer together.

The Marine Mammal Center (TMMC); Marin Headlands; Golden Gate National Recreation Area; Sausalito, CA 94965; Tel. 415-289-7325; www.tmmc.org

> TMMC is a non-profit organization which rescues, rehabilitates, and releases wild, injured marine mammals, does cruelty-free marine research, and offers educational programs.

Maxi Photo; 2301 W. Hwy. 89A; Sedona, AZ 86336; Tel. 520-282-1107

> Maxi Photo is a photographic laboratory featuring one hour color developing, same day black and white developing, aerial photography and a one hour portrait studio.

One World Projects, Inc.; 21 Ellicott Ave.; Batavia, NY 14020; Tel. 716-343-4490; www.oneworldprojects.com

> One World Projects offers a variety of goods, including tagua nut jewelry, produced in cooperation with indigenous peoples and with respect for the environment.

Yolanda Pelayo; PO Box 588; Sedona, AZ 86339; Tel. 520-204-9049; www.gritzka.com/yolanda

>Yolanda is a talented photographer who has photographed people and products all around the world.

The Red Victorian Bed, Breakfast & Art; 1665 Haight St.; San Francisco, CA 94117; Tel. 415-864-1978; www.redvic.com

>The Red Victorian is a friendly, colorful, small hotel featuring vegetarian breakfasts, non-animal tested toiletries and a dedication to promoting peace between all life forms.

Rob Robb, Ph.D.—Adventures in Self-Mastery; 316 Mid Valley Center Ste. 269; Carmel, CA 93923; Tel. 800-887-7622, or 408-624-5966

>Rob is an extraordinary psychic who sees the aura colors of people (and animals), interprets messages from angels and gives consultations to improve people's lives.

Amanda Sehic; www.sehic.net/amanda

>Amanda, a member of Models with Conscience, is also a talented artist. One of her exquisite drawings appears in this book. You may view others on her personal Website.

Russell Tanoue Photographer; 350 Ward Ave. Ste 106; Honolulu, HI 96814; Tel. 808-846-5826; www.russelltanoue.com

>Russell is a professional photographer who specializes in capturing the beauty of both male and female models and actors of all ages, races and ethnicities.

Resources 137

Gary connecting with Cricket and Cera, burros at Hacienda de los Milagros Animal Sanctuary. T-shirt by EnviroLink. Photo by Yolanda Pelayo.

The San Francisco Society for the Prevention of Cruelty to Animals (SF/SPCA); 2500 16th St.; San Francisco, CA; Tel. 415-554-3000; www.sfspca.org
> The SF/SPCA is a non-profit corporation which led San Francisco's implementation of a no-kill animal control policy, and runs an exemplary, cageless animal adoption center.

Vegetarian Shoes; 12 Gardner Street; Brighton, East Sussex BN1 1UP England; Tel/Fax +01273 691913; www.vegetarian-shoes.co.uk
> Vegetarian Shoes is an excellent source for animal-friendly apparel, including fine non-leather shoes, boots, jackets, pants, gloves, wallets, belts and more!

A beautiful woman with a beautiful message—Amanda in the Red Victorian's art gallery, holding a peace poster by Sami Sunchild. Jacket, blouse and pants by Esprit. Photo by Russell Tanoue, make-up by Jessica Campo, hair by Danica Winters.

Models with Conscience members holding plush animals at The Marine Mammal Center. Upper row, Left to Right: Suzanne, Julie, Daniel, myself (in a faux fur jacket by Esprit); middle row: Jacqueline, Amanda, Gary; lower row: Petal and Crystal. Photo by Russell Tanoue, make-up by Jessica Campo, hair by Danica Winters.

In the end, only kindness matters.
—Jewel Kilcher